Saline District Library

Provided by
a grant
from the

CARL F. SCHRANDT
ENDOWMENT
FUND

the climate
REVEALED

the climate REVEALED

William J Burroughs

CAMBRIDGE
UNIVERSITY PRESS

THE CLIMATE REVEALED
William J Burroughs

PUBLISHED BY THE PRESS SYNDICATE
OF THE UNIVERSITY OF CAMBRIDGE
The Pitt Building, Trumpington Street,
Cambridge CB2 1RP United Kingdom

CAMBRIDGE UNIVERSITY PRESS
40 West 20th Street, New York, NY 10011-
4211, USA http://www.cup.org

First published in 1999 by Mitchell Beazley,
an imprint of Octopus Publishing Group Ltd
2–4 Heron Quays, London, EI4 4JP
Copyright © Octopus Publishing Group Ltd 1999

First published in the United States in 1999.
Library of Congress Cataloguing in Publication
data available

ISBN 0 521 77081 5

Typeset in Goudy, Helvetica and Helvetica Neue

Printed in China by Toppan Printing Company Ltd

CONTENTS

INTRODUCTION

THE CLIMATE REVEALED EXPLORES THE VARIETY AND COMPLEXITY OF ALL THE INTERCONNECTED ELEMENTS OF THE EARTH'S CLIMATE.

The climate plays a fundamental role in all our lives. Whatever we do and wherever we live in the world, our daily routine is built around what the weather conditions are like. When the unexpected happens, it can be a minor inconvenience for some people and a matter of life and death for others. Furthermore, in today's increasingly mobile and travel-oriented global culture, long journeys to far-flung parts of the world are becoming more and more common. It is vital in such circumstances that we understand and appreciate what the climate of our destination is likely to be, and prepare accordingly.

This book explores the many aspects of the climate and the impact each can have on our lives. To do this it divides the world up into broad climatic zones: for example, tropical regions, temperate areas and the polar climes. Each chapter opens with an introduction outlining the climate of each zone and then goes on to explore extremes of weather, the vegetation and wildlife and how these relate to environmental issues and human health. All the individual strands are linked together to enable readers to see how the parts make up the whole. From the food we eat and the plants growing in our gardens to the attractions and risks of leisure activities, such as hiking and skiing, all are affected by the climate. Particular emphasis is given to the influence of the climate on the human body and health. In this way, it is possible to provide new insights into how the weather is relevant to almost every aspect of our lives.

Below: The Dust Bowl years were caused by intensely hot dry spells of weather that struck the Great Plains in the USA during the 1930s. They had a devastating impact on the agriculture of the region.

To set these specific themes in context, the opening chapter provides a general introduction to the systems that control the global climate. Complicated physical processes of the atmosphere and oceans are explained and colour illustrations illuminate the most complex issues. In the following chapter there is a brief guide to how meteorologists measure the climate, what they have been able to fathom about climatic conditions of the past and how they predict weather in the future. Central to the question of climate change is an understanding of how national weather services model the climate and predict the future, and how the intricate network of connections between different features of the climate is considered from many angles. This emphasis on the interconnection of the global climate reflects our growing understanding of the behaviour of the whole system. Every two-page discussion of an issue has a comprehensive list of cross references to the other pages that cover related subjects.

Particularly dramatic elements of the Earth's climate, such as avalanches and hurricanes, are highlighted through the book in special feature spreads. These capture what it is like to live through such extreme weather events and explain how they can touch people's lives, sometimes with devastating or disastrous consequences.

Lurking behind these issues is the all-pervading question of global warming and what this means for the future of the planet. Any assessment of the extent to which current events are a consequence of human activities requires careful handling. For this reason, particular attention is devoted to putting recent extremes into the context of past events (for example, comparing recent American droughts with the Dust Bowl years that shattered the lives of farmers in the Great Plains during the 1930s, and assessing the claims about the changing incidence of hurricanes and their economic consequences). Only by forming a balanced picture of how various aspects of our climate are changing, and by deciding whether we can trust predictions of future changes, can we participate in the debate about what price should be paid to reduce the impact that greenhouse gases are having in the atmosphere.

By linking these numerous diverse threads of the climate debate and their influence on many different aspects of our lives, the objective of this book is to tease out just how subtle and all-enveloping the connections are. In this way, the readers can obtain a better feel for the fascinating, intricate complexity of the Earth's climate, and discover just how much the daily life of every one of us is governed by its mercurial behaviour.

HOW TO USE THIS BOOK

The Climate Revealed explores the awe-inspiring diversity and compexity of the Earth's climate. The author opens with an introduction to fundamental climatic processes and looks at the methods used, both past and present, to record the climate. He then describes the Earth's various climatic zones, highlighting what makes each one unique. He focuses on the impact the climate has had on our lives and the influence that we, in turn, have had on the climate. Particularly extreme or dramatic weather events are given special emphasis in feature articles. Finally, the author looks at what may happen to the climate over the next 100 years.

Average temperature and rainfall graphs for places in each area appear on the introductory pages of each climatic zone.

Locator maps show you the areas that are covered by each climatic zone.

▼ **GENERAL FEATURES**

Throughout the book, the author uses clear, jargon-free language to explain complex issues. The accompanying photographs and illustrations illuminate particularly important or complicated aspects of the climate.

Graphs showing long-term trends in various climatic factors appear throughout the book. Each line is an individual reading and the black curve is an average of all these readings.

CLIMATE OF THE MOUNTAINS

MOUNTAINS ARE UNIQUE BECAUSE THEY GENERATE THEIR OWN WEATHER AND SO TRANSCEND OUR USUAL INTERPRETATION OF CLIMATIC ZONES.

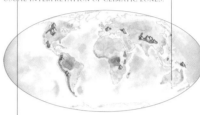

up to which snow melts away in summer. And they must have enough snow to produce distinctive glacial landforms. Often termed 'high mountains', peaks meeting these criteria vary in height according to their location: ranging from 1500m (5000ft) in the Alps to around 4500m (15,000ft) in the Andes on the equator.

THE BASICS OF MOUNTAIN WEATHER
The most basic feature of the mountain climate is that the higher it is, the colder it becomes. On average the temperature drops by 6.5°C for each kilometre (or 3.6°F every 1000ft) in altitude. This decline in temperature with altitude is known as the lapse rate and can vary according to weather conditions: moist warm air cools more slowly than cold dry air because it releases heat as water vapour condenses out. The lapse rate in tropical mountains is less rapid than that at higher latitudes.

Another obvious feature of the climate of mountains is that the amount of rain and snow increases with altitude. In mid-latitudes the amount of precipitation doubles between sea level and about 1500m (5000ft), and then doubles again by the time it reaches 3000m (10,000ft). The heaviest precipitation occurs on the windward side of mountain ranges, falling off sharply on the sheltered leeward side to create an area of low rainfall known as a rain shadow. This happens because, when weather systems pass over mountains, the air is forced to rise and, as it cools, rain and snow condense out. By the time the air reaches the leeward side of the mountain, therefore, there is little moisture remaining in the air.

Below: Hidden Peak in the Karakoram range in Pakistan. The mountain is seen from the south-west, near the base camp situated on the Abruzzi Glacier.

SONNBLICK, AUSTRIA

WHAT ARE MOUNTAINS?
It may seem odd, but we must start by defining what we mean by mountains: peaks that are significantly higher than the surrounding countryside. They must have a clear upper tree-line, where forests thin out, and snow-line, the point

Millions of us visit mountainous regions every year to enjoy winter sports. However, mountain weather can be difficult to predict and there are a number of challenges to be faced: torrential rain and snow, ferociously strong winds, bitterly cold temperatures and a high risk of sunburn from the Sun's harmful ultraviolet rays.

LEH, PAKISTAN

□ Average monthly temperature
／ Monthly precipitation

12 Electromagnetic spectrum
20 How do clouds form?

82 MOUNTAIN REGIONS

HISTORICAL RECORDS

HISTORICAL RECORDS ARE IMPORTANT TO OUR UNDERSTANDING OF PAST CLIMATIC CHANGE – THEY ALSO ADD COLOUR AND CONTEXT TO PROXY DATA.

Historical records, such as personal diaries and agricultural records of the condition and price of crops, are useful sources of information about past weather, especially where instrumental observations do not exist.

WHEAT PRICES
In Europe and China detailed reports of many annual events going as far back as the Middle Ages provide valuable insights into the weather conditions at the time. This is particularly useful where historical records overlap with instrumental observations and so can be checked against reliable measurements. Their frequent dependence on agricultural activity is, however, a limitation. For example, the price of wheat is a good indicator of the abundance of the harvest because bumper years led to low prices and a dearth produced high prices. But, wheat harvests only relate to the summer period and depend on both temperature and rainfall. This means that, while the extreme years stand out, many of the fluctuations cannot be attributed to a specific factor.

WINE HARVESTS
The wine-harvest date, when grapes were picked for wine-making, is a more reliable measurement of the temperature during the growing season in the northern hemisphere (April to September). The best-known record of wine-harvest dates has been built up from documents kept in parts of northern France where there is a centuries-old

tradition of villages and towns declaring the local vineyards formally ready for harvesting. These records provide a good guide to summer temperatures between 1484 and 1879. Long-term fluctuations do, however, have to be viewed with caution. For instance, a period of late harvest dates in the late 18th century and early 19th century does not tally with available instrumental records for the period, and has more to do with changing fashions in the sweetness of wine: the later the harvest, the sweeter the wine.

FILLING THE GAPS
The principal limitation of many other historical records is that frequently they only note exceptional events. This means that many records are often exceedingly patchy, as they cover only a few years in any century. A good example is the record of polar bears coming ashore from ice floes around Iceland. On the face of it, accounts of polar bears in Iceland suggest that the Arctic pack-ice extended further south in the 13th and 14th centuries. But there are big gaps in the record, which mean we may be attaching too much importance to the fact that these rare events became a little more common after about 1300.

The way round this limitation is to find more records that may tell us about the climate at the time. An examination of all the extremes recorded can make it possible to build up a more complete picture of changes. An example of this is the work of Christian Pfister, at the University of Berne, Switzerland, on the changes in temperature and rainfall in Switzerland for each season since the early 16th century (see below). He combined written records of snowfall, freezing of rivers and floods with flowering dates of trees and plants, cereal and wine-harvest dates and yields, to provide a year-round measure of the climate.

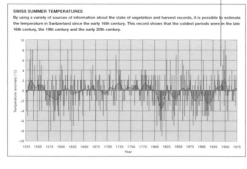

Above: Harvesting was always a time of maximum effort when the weather was good – any deterioration in conditions could spoil the precious grain.

SWISS SUMMER TEMPERATURES
By using a variety of sources of information about the state of vegetation and harvest records, it is possible to estimate the temperature in Switzerland since the early 16th century. This record shows that the coldest periods were in the late 16th century, the 19th century and the early 20th century.

in rainfall during the summer across this vast country. The references are extensive because, every year during the last five centuries, the Chinese authorities have graded the droughts and floods for over one hundred locations in the eastern part of the country.

In many years, the pattern of rainfall is fragmented, with plentiful rain in some parts of the country and drought in others. For example, in 1560, both the north and the south had drought while the Yangtze basin was flooded. The records include such items as: in Datong men were eating men, in Shijiazhuang there was no rain in spring or summer and in Beijing there was a locust plague. In contrast, there were overwhelming floods in some other regions including Nanking and Yichang.

What emerges from this analysis is that during the summer half of the year the country is usually in the grip of one of about half-a-dozen basic weather regimes, such as droughts or floods. However, the records show no particular sequence for the occurrence of these regimes, or any marked change in the incidence of one or more of them over time. So, all in all, they suggest that the climate of this region has experienced the same considerable fluctuations for at least the last five centuries, and probably even longer.

As a rule, historical records can rarely be used to build up a real picture of climate change. Where the records are few and far between, great care has to be taken in not

attaching too much importance to fluctuations in rare events. Only when there are frequent observations over a significant period can these records be used on their own to draw conclusions on climate change. Otherwise, their main value, in conjunction with proxy data, is that they provide background detail in assessing the social impact of climate variations and in building up a more complete picture of any change. Even if not strictly reliable in their own right, historical records do add colour to instrumental measurements and proxy data, and show how climate change has affected society and people.

Proxy records 40–1

Past wine harvests 103

38 CLIMATE RECORDS

HISTORICAL RECORDS 39

Stunning photographs capture the dramatic and potentially catastrophic nature of the Earth's climate.

THE IMPACT OF VOLCANOES

There are few natural phenomena as dramatic, and for humans as potentially devastating, as a volcanic eruption. Molten lava is the immediate danger but other emissions have longer-lasting effects.

Benjamin Franklin was the first to suggest that volcanoes could affect the climate. He proposed that the bitter winter of 1783–4 in northern Europe was due to the dust cloud from the eruption of Laki in Iceland in July 1783, which dimmed the Paris sky for months.

Dust veils and sulphur

Explosive volcanic eruptions can inject vast amounts of dust and sulphur dioxide into the atmosphere. Both have climatic repercussions. Sulphur dioxide has the greatest impact on the climate, as when propelled into the upper atmosphere it is converted into tiny sulphuric acid droplets. At altitudes of 15–30km (50,000–100,000ft), where there is no significant vertical air motion, these minute particles can remain suspended for several years and spread around the globe. In addition, solid particles form a dust veil in the upper atmosphere. Both the droplets and the dust veil absorb sunlight, therefore heating the stratosphere but cooling lower levels by reducing the amount of solar radiation reaching the Earth's surface.

Measuring volcanic eruptions

Analysis of how past volcanic activity affected the climate is problematic because worldwide historical records were not kept until the late 19th century and so it is hard to be certain of global changes. In the 20th century, there were no significant eruptions until 1963 when Agung in Bali erupted. But since 1980, three major but very different volcanic events have given climatologists the opportunity to discover much more.

The first of these was when Mount St Helens in the USA erupted in 1980. Contrary to popular belief, this had no notable climatic impact because it blew out sideways and did not inject much dust high into the stratosphere. More significantly, its plume was low in sulphur. Although it only had a limited effect in terms of cooling, it provided valuable insights into the role of sulphur.

The second eruption, of El Chichón in Mexico in 1982, though a fairly small volcano, was high in sulphur content. Satellite measurements of the stratospheric aerosol cloud, made up of tiny sulphuric acid droplets, confirmed the importance of sulphur. Finally, the most important eruption climatically was Mount Pinatubo in the Philippines in 1991, which injected around 20 billion kg (20 million tons) of sulphur compounds into the stratosphere.

Scientists used these eruptions to test climatic theories by making atmospheric measurements and using computer models of the climate. Accurate measurements of the temperature of the stratosphere by satellites showed how volcanic dust clouds warmed the upper atmosphere. The surface-temperature records revealed a cooling in the lower atmosphere, as predicted by climate computer models. These observations also confirmed that the effects of a single eruption can last for around two to three years.

Evidence from early volcanoes

Recent studies of Antarctic and Greenland ice cores and tree rings around the northern hemisphere have produced a more detailed picture of the link between volcanoes and the climate. These identify which eruptions had the greatest impact on the atmosphere, and the influence they had on weather patterns. In the northern hemisphere the greatest cooling effect is during the summer; winters are, if anything, warmed by the change in atmospheric circulation caused by eruptions.

In the last 600 years or so the coldest summer was in 1601, and appears to have been caused by a major eruption in Peru the year before. The so-called 'year without a summer' of 1816 was the second coldest on record and was probably due to the massive eruption of Tambora in Indonesia in 1815, which injected five to 10 times more material into the stratosphere than Pinatubo would later do. Exceptionally late frosts destroyed crops in New England, and French vineyards had their latest wine harvest for at least five centuries.

The short-lived impact of volcanoes suggests they can only trigger lasting change if they coincide with other factors that also stimulate cooling. For example, the largest eruption in the last million years – Toba in Sumatra, 73,000 years ago, which was at least five times the size of Tambora – coincided with a rapid cooling during the development of the last ice age. This eruption could have produced sufficient cooling for perennial snow cover to form for a number of years in, say, northern Canada. At a time when the Earth was already slipping into a cold phase, this snow cover, with its additional cooling effect due to reflection of more sunlight into space, may have tipped the balance.

When Mount Pinatubo erupted in June 1991 after being dormant for more than 600 years, it took not only scientists by surprise but also its Filipino neighbours, whose life it touched in a dramatic way. The first inkling of Pinatubo's fury was a deep rumble, which sounded like thunder coming from inside the Earth. Then a dark angry mushroom-shaped cloud appeared in the sky and rained sand over the countryside and towns for miles around. After three days the rumbling became more furious and the sky turned black, as dark as night. By a cruel twist of fate, a typhoon struck at the same time, and the rain mixed with the volcanic ash and fell from the sky as mud. The sulphur in the air, the overwhelming stench and the huge fiery displays of lightning were an apocalyptic combination. Then the earthquakes began, bringing buildings crashing to the ground. Over 300 people were killed by the eruption and it destroyed the homes and livelihoods of thousands more.

Above: Satellite images show how the dust cloud from Mount Pinatubo was swept around the world in just 10 days by winds in the stratosphere.

Left: The eruption of Mount Pinatubo in 1991 provided the clearest evidence yet of how volcanoes affect the climate.

asses over moun ss, which leads to f the range and the le. In going around celerates through swirls down valleys, creating uble variations in snowfall and high snowfall (Schneewinkel) s (Chinook or Föhn).

other parts of the world, from the equatorial Andes and the Himalayas to Norway, Iceland and Spitzbergen, there is enough snow to maintain sizeable ice fields and glaciers. These icy regions are, however, under threat from the consequences of global warming.

IN SNOW

d snowfall, together with the ow long snow lies and whether ers are concerned, a rule of over occurs when the average w –3°C (27°F). Places that are avier snowfall build up a base aws. The level of permanent ependent on the temperature fall and sunshine. urs on the windward slopes of influenced by mid-latitude the Cascade Mountains in Southern Alps of New Zealand all ever recorded was 31.1m ger Station, on Mount Rainier, n altitude of 1655m (5428ft) 2 the snow had compressed ft).

SUNSHINE

Mountainous regions often get much more sunshine than the lowlands around them. For instance, during the winter, the uplands in the Alps get far more sunshine than lower down because of the low-level clouds. In the winter, temperatures in the valleys fall to lower values and the air close to the snow-covered slopes cools and slips downhill to lowlands. This creates a temperature inversion, which traps pollutants and maintains fog and low cloud in the valleys. Conversely, in summer, much of the snow melts away and the lowlands receive more sunshine. This causes convection and warm air flows up the mountain to form clouds around the mountain tops. Overall, the annual figures are not noticeably different but there are serious implications for recreational activities, especially skiing. The other major consequence of sunshine at higher altitudes is the increased level of ultraviolet radiation that is present, which damages the skin.

▲ FEATURE ARTICLES

Dramatic, topical and controversial weather events are given special attention in feature articles. From volcanoes and hurricanes to avalanches and El Niño, these present an objective view of issues that are scarcely ever out of the headlines.

Satellite images are used throughout the book to present data in an illuminating and easy-to-understand way.

Down the right-hand side of each feature article is an account that highlights the impact that the dramatic weather event has had on humankind.

▲ INTRODUCTORY PAGES

Each of the chapters that explore the climatic zones opens with introductory pages giving an account of what the weather in each area is like and an outline of the weather systems that play an important part in the climate.

Colour-coded cross references lead you directly to the page where related issues are explained in detail.

Each chapter is colour-coded to allow you to find your way around the book easily.

▼ CHAPTER COLOUR CODING

Introduction
Climate in motion
Climate records
Polar regions
Tundra and taiga
Mountain regions
Mediterranean regions
Temperate regions
The prairies
The tropics
Desert regions
The future

CLIMATE IN MOTION

THE EARTH IN SPACE

THE SUN IS THE DRIVING FORCE OF OUR SOLAR SYSTEM: IT SUSTAINS OUR LIFE ON EARTH AND CONTROLS ALL CLIMATIC PROCESSES.

WHAT IS RADIATION?
Radiation is the transfer of energy in the form of electromagnetic waves of varying wavelengths, which are denoted in micrometres. The light that we see – visible light – is a form of energy and is just one part of the electromagnetic spectrum. Other types of energy are used widely in science, technology and medicine: X-rays, radio waves, ultraviolet light and gamma rays. Incoming solar radiation is concentrated in ultraviolet, visible and infrared wavelengths, shown in yellow on the diagram below. Outgoing terrestrial radiation is concentrated in the mid-infrared wavelength and is shown in red. The smooth outer curves represent a simplification of the complex processes of radiation emission.

All aspects of the Earth's climate are governed by the energy received from the Sun. The Earth also radiates energy back out into space. Over time there is a balance between the energy coming in and going out: the amount of solar radiation received by the Earth is equal to the amount of heat radiated into space.

PLANET EARTH

The Earth is the only planet in our solar system on which life is known to exist. The other planets are extremely inhospitable: some have poisonous atmospheres, some have no solid surfaces, those close to the Sun burn in enormously high temperatures and those furthest away are frozen far below zero. The Earth is unique because its position in relation to the Sun allows it to maintain an equable balance of temperature, atmosphere and pressure.

FUELLED BY THE SUN

Many of the Earth's climatic processes are powered by the Sun as it heats the surface of the lands and the oceans. The water cycle is driven by solar energy causing water to evaporate and form clouds. This energy is also vital in maintaining the biosphere because it is necessary for photosynthesis, which in turn enables plants to grow.

Most of the energy reaches the Earth at the equator and is transported towards the poles by the atmosphere and the oceans. These transporters of heat define the climatic zones around the Earth: from the mild temperate regions of northern Europe to the icy conditions at the poles and the huge extremes of temperature that characterize the North American prairies and Siberia.

INCOMING ENERGY

Of solar radiation that reaches the Earth, 9 per cent is concentrated in the ultraviolet wavelength band, 45 per cent is visible light (where it is most intense) and the remainder is at longer wavelengths. Part of this radiation is absorbed before it reaches the Earth's surface: ultraviolet radiation is absorbed by oxygen and ozone high in the atmosphere, while infrared radiation is absorbed by water vapour and carbon dioxide at lower atmospheric levels.

OUTGOING ENERGY

The Earth's upwelling energy, terrestrial radiation, is radiated out from both its surface and atmosphere. It is emitted in the mid-infrared wavelengths and, the warmer the surface or atmosphere, the more radiation is emitted.

The amount of energy radiated by the atmosphere is controlled not by the main atmospheric gases (oxygen and nitrogen), but by certain trace gases, notably water vapour, carbon dioxide and ozone. Each of these interacts with infrared radiation in its own way, absorbing and re-emitting radiation into the atmosphere. The trace gases can also affect the level in the atmosphere at which energy is radiated out to space. If there is a higher proportion of a particularly absorbent gas in the lower atmosphere, then it will absorb and radiate from there rather than from higher levels, thus influencing the behaviour of the Earth's atmosphere as a whole.

RADIATION AND ABSORPTION
Incoming solar radiation (incoming energy) is both reflected and absorbed by clouds, the atmosphere and the Earth's surface. The atmosphere absorbs one-fifth of this radiation and the surface nearly half of it. The remainder is reflected straight back into space. Some of the absorbed solar energy generates convection in the atmosphere and also stimulates the evaporation of water. This energy flow is combined with the emission of terrestrial radiation (outgoing energy) to drive the climate system. Overall the net outgoing terrestrial radiation from the Earth's surface and atmosphere balances the amount of absorbed incoming solar radiation.

OUTGOING ENERGY (REFLECTION)
31 per cent of solar energy is reflected back out to space by clouds and the Earth's surface. Deserts and snow cover are particularly reflective.

Gamma rays	Short wavelengths
X-rays	
Ultraviolet	
Visible	
Infrared	
Incoming energy	
Outgoing energy	
Microwaves	
Radio waves	
	Long wavelengths

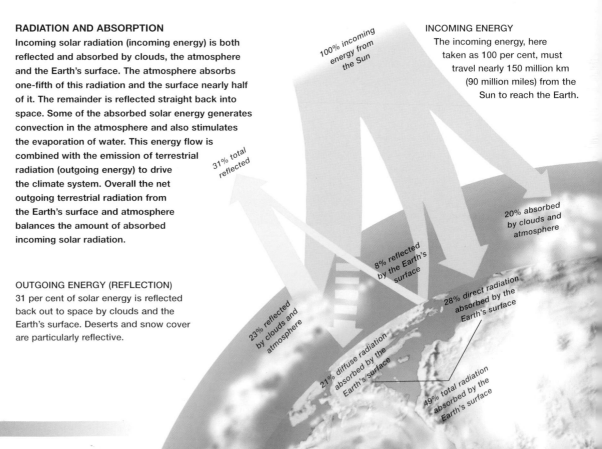

100% incoming energy from the Sun

INCOMING ENERGY
The incoming energy, here taken as 100 per cent, must travel nearly 150 million km (90 million miles) from the Sun to reach the Earth.

31% total reflected

20% absorbed by clouds and atmosphere

8% reflected by the Earth's surface

28% direct radiation absorbed by the Earth's surface

23% reflected by clouds and atmosphere

21% diffuse radiation absorbed by the Earth's surface

49% total radiation absorbed by the Earth's surface

WHY DO WE HAVE SEASONS?

The Earth rotates on its axis every 24 hours giving us day and night, and orbits around the Sun every 365 days giving us seasons. The seasonal cycle depends on the 23° tilt of the axis – the Earth always points in the same direction in space and, as it orbits, each hemisphere is tilted away from and towards the Sun producing summer and winter. When the Sun is directly over the Tropic of Capricorn in late December, the southern hemisphere has its longest day (summer solstice) and the northern hemisphere has its shortest day (winter solstice). At this time, the North Pole is hidden from the Sun and has continual night, and the South Pole is bathed in thin sunshine 24 hours a day. In late June, when the Sun is directly over the Tropic of Cancer, the solstices are reversed. In between these two positions, the Earth inclines less towards or away from the Sun and the two hemispheres have spring and autumn respectively. When the Sun is directly over the equator, at the March and September equinoxes, both hemispheres have equal day and night.

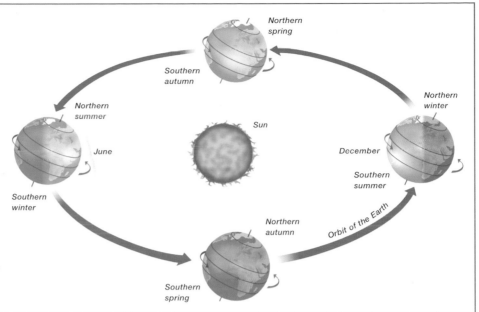

THE ATMOSPHERIC BLANKET

The Earth's atmosphere consists of nitrogen, oxygen, water vapour, argon and carbon dioxide, and was formed early in our planet's life as the gases escaped from its interior by way of volcanic eruptions. Since then its composition has been determined by the development of life, in particular the evolution of plants that absorb carbon dioxide and emit oxygen, necessary for the survival of animal life.

The atmosphere is vital to the existence of life on the Earth. It has two effects: firstly it protects organisms from the Sun's harmful rays by absorbing them before they reach the Earth's surface; and secondly it acts as a blanket or greenhouse, absorbing the energy emitted by the Earth and radiating it back towards the surface (see below).

The gases that trap the heat are known as radiatively active gases: water vapour contributes the most to the overall warming of the planet, followed by carbon dioxide and then ozone. Increases in the levels of these gases will have a warming influence on the Earth's climate, by enhancing the greenhouse effect.

The influence of carbon dioxide and ozone highlights the potential impact that human activities can have on the Earth's climate by altering the concentrations of these gases in the atmosphere. Other products of human activities, which include methane, oxides of nitrogen, sulphur dioxide and chlorofluorocarbons (CFCs), also modify the radiative properties of the atmosphere and increase the greenhouse effect.

69% outgoing energy (emission)

48% emission from atmosphere

7% energy released by convection

23% energy released by evaporation and transpiration

9% emitted by clouds

12% direct energy from surface

114% emitted from the Earth's surface

Energy absorbed and re-emitted by greenhouse gases

95% re-emitted into the atmosphere

OUTGOING ENERGY (EMISSION)

69 per cent of the energy received from the Sun is radiated back out to space by the Earth's surface, clouds and atmosphere. The rest is absorbed and fuels the Earth's climate system.

THE GREENHOUSE EFFECT

Radiatively active gases are usually described as 'greenhouse gases' because they trap some of the Earth's outgoing radiation and reradiate it back towards the surface. This is vital in maintaining a climate warm enough to sustain life. But if the levels of these gases increase, then the so-called greenhouse effect increases global warming and can become a problem. The 114 per cent shows how the Earth emits more energy than it receives from the Sun because energy is recycled by the process of absorption and re-radiation into the atmosphere – only 12 per cent escapes directly to space.

THE ATMOSPHERE IN MOTION

OUR ATMOSPHERE IS NOT A STATIC LAYER OF AIR BUT INSTEAD A DYNAMIC SWIRLING ENGINE THAT CARRIES HEAT FROM THE SUN AROUND THE GLOBE.

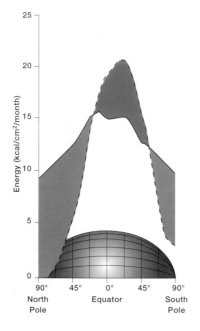

GLOBAL RADIATION BALANCE
Most incoming solar radiation from October to March (here shown as a dotted line) enters at the equator and at low latitudes, and is transported polewards before being re-emitted as outgoing terrestrial radiation (here shown as a solid line). The net gain of energy is shown in red and net loss is shown in blue.

Right: This false-colour image of the Earth from space, recorded on 2 September 1994, shows a variety of weather systems over the eastern Pacific, North and South America and the western Atlantic ocean. Over the equatorial regions there are bands of cloud caused by the ITCZ.

The movement of the atmosphere and oceans plays a vital part in the transfer of energy around the globe. Most solar energy enters at the equator, and sets the atmospheric-energy transfer engine in motion. Hot moist air rises and moves towards the poles in a series of cell-shaped patterns (see right).

THE VIEW FROM SPACE

Astronauts looking down to the Earth from space get a unique view of our atmosphere in motion: an endless succession of transient eddies swirling erratically. However, if you look at them for long enough, clear patterns begin to emerge. The most obvious is the huge pulse of the annual temperature cycle of the seasons that appears to move from hemisphere to hemisphere. When monitored with heat-sensing equipment, this looks like a great blush spreading across the face of the Earth during the year.

In the equatorial regions the planet is girdled by a region of intense convective activity, known as the intertropical convergence zone (ITCZ). During the summer in the northern hemisphere this develops into the monsoon – the rainy season that spreads over the Indian subcontinent. Then, during late summer and early autumn, activity in the ITCZ erupts into more organized outbursts. These result in tropical cyclones that emerge in the easterly flow of air and then, as they intensify, swerve northwards before merging into the high-latitude westerly flow. To the north and south of the tropical hive of activity the continental deserts, such as the Sahara, mark regions of sinking air and low rainfall. Beyond this, the westerly conveyor belt of mid-latitude depressions swirls endlessly towards the poles.

In the southern hemisphere, this procession of air continues unabated throughout the entire year with only a modest shift to higher latitudes taking place during the summer months. In the northern hemisphere, there is a crescendo of activity in the winter half of the year, with a marked lull during the summer because the contrast in temperatures between the tropics and polar regions declines considerably at this time.

THE ATMOSPHERIC MOTION

The large amounts of sunshine received in the equatorial regions drive the atmospheric motion, which transports energy around the entire globe.

EQUATOR TO MID-LATITUDES

Energy enters at the equator and sets the convective activity of the ITCZ in motion because warm moist air rises. By the time it reaches an altitude of 12–15km (7–9 miles), the temperature of the air has fallen to about –80°C (–112°F) and virtually all the moisture has been wrung out of it. It then spreads out and descends at about 20–30° latitude. First interpreted by George Hadley in 1735, this circulation pattern now bears his name (the Hadley Cell). Some of the air flows back to the equator to form the ever-steady flow of the Trade Winds.

MID-LATITUDES TO POLES

The flow from the Hadley Cell is driven polewards and creates the westerlies in the mid-latitudes. Here, the circulation patterns are less well-defined and cold sinking air clashes with warm rising air to form depressions, which characterize the climate of these regions. The depressions raise huge quantities of air, which descend over polar regions to create cold deserts in these areas before flowing back towards the equator to start the process again.

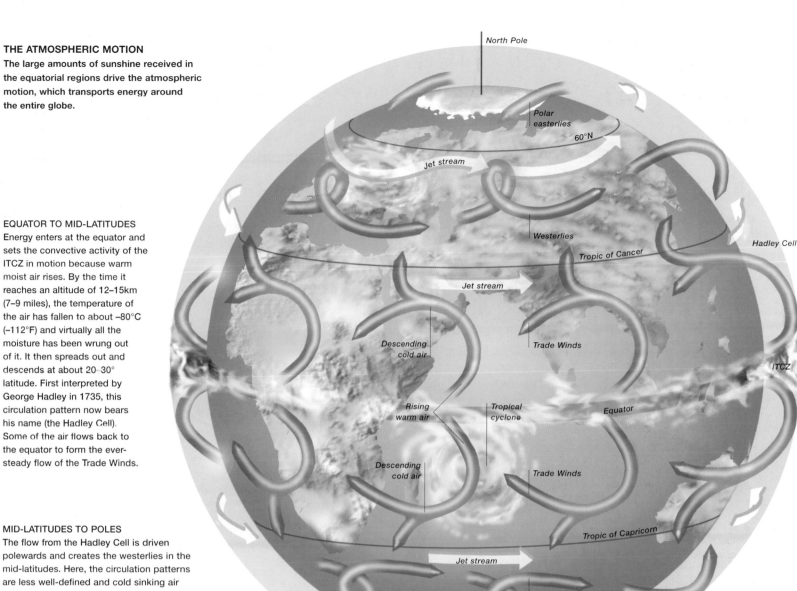

North Pole
Polar easterlies
60°N
Jet stream
Westerlies
Tropic of Cancer
Hadley Cell
Jet stream
Descending cold air
Trade Winds
ITCZ
Rising warm air
Tropical cyclone
Equator
Descending cold air
Trade Winds
Tropic of Capricorn
Jet stream
Jet stream
Westerlies
60°S
Polar easterlies
South Pole
Rotation of Earth

UPPER-ATMOSPHERE WINDS

Weather watchers around the world are now able to recognize distinct patterns of air circulation, which are an essential part of putting together reliable weather forecasts. These circulation patterns are most clearly seen in the upper atmosphere where there are strong winds, known as jet streams. Understanding the behaviour of jet streams is the key to producing long-term forecasts. Air pilots also have to be aware of these winds: flying against or in the same direction as a jet stream can make the journey time much longer or shorter respectively.

Jet streams occur at heights of 9–14km (30,000–45,000ft) and are concentrated between two narrow bands of latitude, one of 40–60°N and S and the other around 30°N and S. They usually travel at speeds of 160–240kph (100–150 mph) but can exceed 450kph (280 mph) in the winter.

THE PATH OF JET STREAMS

Jet streams follow wavy meandering patterns around the Earth. These are determined by the rotation of the globe and by the circulation of the global atmosphere. When poleward-bound warm air rises in a low-pressure system, it produces what is known as a ridge in the jet stream. Conversely, where cool air in a high-pressure system sweeps back to the equator, the descending air produces a trough in the jet stream. The number of troughs and ridges at any one time may vary and is hard to predict.

This interaction between surface weather systems and the jet stream also works in reverse because the upper-atmosphere winds tend to steer the movement of the surface weather features. It is these connections that weather forecasters need to understand in order to produce more accurate predictions.

OCEANS AND CURRENTS

COVERING OVER 70 PER CENT OF THE EARTH'S
SURFACE, THE OCEANS ARE VITAL IN CONTROLLING
OUR CLIMATE AND IN SUSTAINING LIFE.

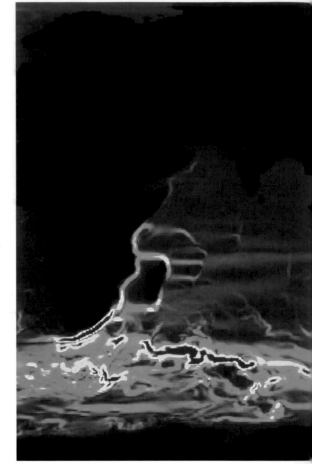

The vast oceans contain a mind-boggling 1.3 billion km³ of water. Not only is this the most fundamental component of the life-sustaining water cycle, but the oceans act in tandem with the atmosphere in transporting energy around the globe from the equator to the poles.

WHY ARE OCEANS SO IMPORTANT?

Oceans play a vital part in controlling global temperatures. The sea heats up and cools down more slowly than land masses. This is important in maintaining moderate air temperatures in coastal areas, because in summer the oceans absorb sunlight at deeper levels and so warm less rapidly than land. The consequence is that maritime regions are spared the extremely cold and hot conditions experienced by continental interiors.

Not only do oceans have a greater heat capacity, they also transport heat most effectively. A global system of both surface and deep-water currents transports energy around the world – this ensures that the tropics do not overheat and carries heat to the colder higher latitudes.

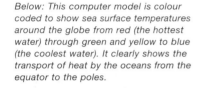

Below: This computer model is colour coded to show sea surface temperatures around the globe from red (the hottest water) through green and yellow to blue (the coolest water). It clearly shows the transport of heat by the oceans from the equator to the poles.

PATTERNS OF CIRCULATION

As an integral part of global energy transport, ocean circulation is driven principally by the wind and modified by the constant rotation of the Earth. Broadly, the oceans in both hemispheres flow westwards out of the tropics and then move to higher latitudes along the western edge of the ocean basins. At latitudes above about 35°N and S the main currents sweep towards the east and carry warm water to higher latitudes. This pattern is most obvious in the North Atlantic in the form of the Gulf Stream/North Atlantic Current and in the North Pacific in the form of the Kuroshio/North Pacific Current.

To compensate for this poleward flow, returning currents of cold water move down the eastern side of the ocean basins. This also happens in the southern hemisphere, but in this case the virtual absence of land between 35 and 60°S means that the flow merges with the strong circumpolar current around Antarctica.

WHAT DO WE KNOW ABOUT OCEANS?

The broad nature of ocean currents has been known for over 100 years to geographers, and was familiar to mariners for centuries before that. The Gulf Stream, for example, which brings mild winters to north-west Europe, was used to determine sailing patterns. The first sketchy plans of ocean currents were put together in the 15th century when European explorers began to take to the seas in search of new territories.

Detailed knowledge of the ocean currents still remains patchy, however, and their structure, the amount of energy they transport and how they change from year to year remain largely a mystery. Oceanographic studies, using ships and buoys, only provide a fragmentary picture but the application of new technology is set to expand our knowledge. Satellite instruments, for example, are able to measure both direction and speed of wind, waves and ocean currents and so are providing new insights into global ocean behaviour.

SURFACE WATERS

Most of our knowledge about the oceans concerns a shallow layer of water on the surface, which does not sink to great depths but moves horizontally and is constantly stirred up by the winds. From the top to the bottom of this mixed layer there is little temperature difference but beneath it is the thermocline – a narrow zone over which there is a

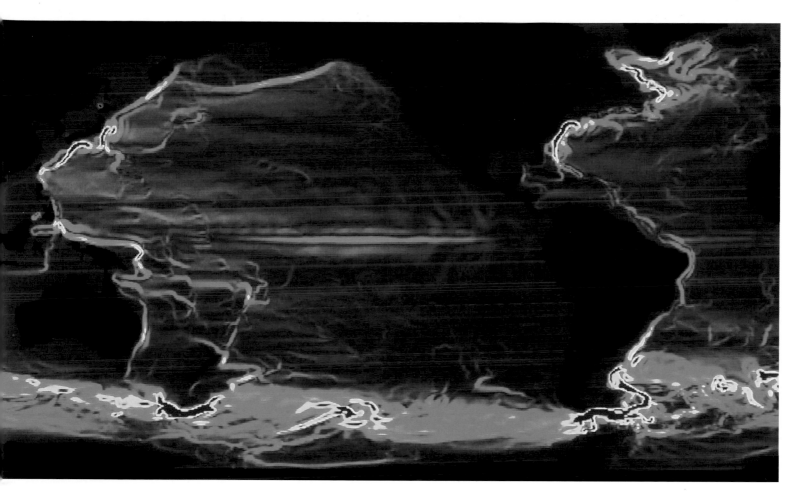

rapid drop in temperature. The thickness of the mixed layer, and hence the depth at which the thermocline occurs, depends on a number of factors: the wind speed, the flow of warmer or colder water horizontally or the upwelling of cold water, and thermal mixing – where the surface waters are heated by the sun or by the passage of warmer or colder air above. The wider horizontal movement of these surface waters is reflected in the broad patterns of the world's ocean currents (see above).

DEEP WATERS

Underneath the seething surface layer of the oceans is a huge body of cold water where there is a much more gradual but equally important set of motions at work. The process driving the deep-water currents is known as thermohaline circulation – water that rises and sinks according to its salinity and density, which in turn are dependent on its temperature. The sinking movement is responsible for the currents that flow along the mountains and valleys of the ocean floor.

In different regions of the world, water rises and sinks to greater and lesser degrees. Where the sinking movement is most significant, the formation of deep waters and bottom waters is the result. Deep waters – defined as water that sinks to middle levels of the major oceans – are formed only around the northern fringes of the Atlantic Ocean. Bottom waters – a colder denser layer underneath the deep

waters – are formed in limited regions near the coast of Antarctica in the Weddell and Ross Seas. It is only in these few regions of the world that sinking waters have a major impact on the climate.

GREAT OCEAN CONVEYOR

Surface and deep-water currents combine to create a global pattern of currents known as the Great Ocean Conveyor (GOC), which transports energy around the globe. This is reflected in the part played by various oceans in transporting heat polewards. The Atlantic carries rather more heat northwards than does the Pacific (the Indian Ocean makes a negligible contribution because it extends so little distance northwards). In the southern hemisphere the Pacific transports roughly twice as much energy southwards as does the Indian Ocean, while the Atlantic overall has a small negative effect carrying energy northwards across the equator.

Any change in the path or strength of the GOC would have a dramatic impact on the climate, and serious repercussions for temperatures around the world. Evidence shows that, during the last Ice age, the GOC followed a very different course, not reaching as far north as it does at present, and that its course changed suddenly over a period of just a few years. A better understanding of the factors influencing the GOC is central to getting to grips with climatic changes being experienced today.

Above: A computer model showing ocean currents and how quickly they move. Areas of red indicate fast-moving water and blue, slow-moving. Several global currents can be seen: the Agulhas (bottom left), the Kiroshio (top middle), the Gulf Stream (top right) and the strong circumpolar current around Antarctica (bottom).

WATER CYCLE AND BIOSPHERE

AN ABUNDANT SUPPLY OF LIQUID WATER SETS THE
EARTH APART FROM THE OTHER PLANETS BECAUSE IT
ENABLES LIFE TO EXIST.

The hydrological cycle – the circulation of water, in liquid, solid or gaseous state, from the oceans into the atmosphere and back to the oceans – is essential to the climate system. All living matter on land and in the oceans depends on a constant supply of water.

WATER

Water is the most abundant liquid on earth. The total amount within the earth–atmosphere system is estimated to be some 1384×10^6 km^3 ; of this amount 97.3 per cent is contained in the oceans, 0.6 per cent is groundwater, 0.02 per cent is contained within rivers and lakes, 2.1 per cent is frozen in ice caps and glaciers (cryosphere) and only 0.001 per cent is in the atmosphere. Nevertheless, water vapour is by far the most important variable constituent of the atmosphere, with a distribution that varies with location and time of year. In addition, on timescales of decades and longer, the proportion of water locked up in the cryosphere can change with important consequences for the climate. During the last Ice age, for example, the cryosphere was three times as large as it is now.

At any one time the average water content of the atmosphere is sufficient to produce a uniform cover of about 25mm (1in) of precipitation over the globe – the equivalent of some 10 days of rainfall. This water is part of a continual recycling process between the oceans, land and atmosphere. It also plays a major part in the energy transport of the climate. Ice absorbs a great deal of energy in the process of melting (latent heat of fusion) and water needs even

more energy to evaporate (latent heat of vaporization). On the other hand, the process of precipitation as rain or snow releases large amounts of energy.

PASSING THROUGH THE CYCLE

The water cycle is the product of all four climatic elements: water, air, heat and land. Very simply, the Sun's warmth causes water to evaporate from the sea and land and, as water vapour rises, it cools and condenses to form clouds. A number of the features of the water cycle have an effect on the climate: how much water passes into the atmosphere, the nature and shape of the clouds it forms and how quickly it precipitates down again.

Over oceans, the temperature of the surface is a major factor in the formation of water vapour, with wind speed playing a secondary role. Over land, the level of soil moisture controls how much water passes into the atmosphere, but the presence of living organisms complicates matters: water evaporates from the soil and transpires from plants (evapotranspiration). These processes depend not only on soil moisture, but also on air and soil temperature, and wind speed. The levels of water vapour in the atmosphere affect how much energy is absorbed and radiated out to space, while its condensation to form clouds is also critical.

WATER AND CARBON CYCLES
The climate system involves two major continual cycles. The first is the evaporation of water from the oceans to form clouds, which then precipitate either back into the oceans or on to the land from where the water makes its way back to the oceans. The other is the carbon cycle – carbon dioxide in the atmosphere is absorbed by vegetation, which produces oxygen. When living systems die, they decay by absorbing oxygen and releasing carbon dioxide into the atmosphere. When there is no ready supply of oxygen they still decay, but form methane and other organic compounds.

Precipitation occurs when water vapour condenses

Water vapour passes through the atmosphere

Oxygen

Carbon dioxide

Carbon dioxide

During photosynthesis, vegetation absorbs carbon dioxide and emits oxygen into the atmosphere

Water evaporates from vegetation to form clouds

Water evaporates from lakes to form clouds

Water from the soil flows back to the oceans

THE ROLE OF THE BIOSPHERE

The role of plants on land is only part of the story. The totality of living matter, both on land and in the oceans (the biosphere) can influence the climate in a variety of ways. The most important of these is the control exerted by the biosphere over certain greenhouse gases. In particular, through photosynthesis, it acts as the fundamental control over the level of carbon dioxide that is present in the atmosphere throughout the year. During the northern hemisphere growing season the biosphere draws down carbon dioxide from the atmosphere, only to release much of it during the winter when leaves and other dead vegetation start to decay. Similarly, methane – an important greenhouse gas – is formed by the anaerobic decay of vegetation inside the ground.

The potential of the biosphere to absorb additional carbon dioxide is an important factor in considering climate change. It can temporarily remove a proportion of any additional carbon dioxide injected into the atmosphere. The higher the carbon dioxide levels, the greater the productivity of the biosphere, which slows down the build-up of carbon dioxide in the atmosphere. This negative-feedback mechanism had a major impact on past climates and can delay some of the consequences of carbon dioxide emissions due to the combustion of fossil fuels.

CREATION OF PARTICULATES

The biosphere also produces particulates: solid particles and liquid droplets, which play a part in the formation of clouds. On land the emission of many volatile organic substances, such as terpenes from fir trees, leads to the formation of particulates and to natural photochemical smogs in hot sunny conditions in forested areas.

In the oceans there is an analogous process: phytoplankton (algae) produce dimethylsulphide (DMS) in order to maintain their osmotic balance with sea water. DMS escapes into the atmosphere, where much of it is converted into sulphate particulates, which help to form clouds. Changes in sea-surface temperatures may lead to more algae and hence alter the amount of DMS released into the atmosphere. If this forms more clouds, it could act as a negative-feedback mechanism to stabilize the climate.

Above: Summer haze over the Great Smoky Mountains in the eastern USA is a good example of a natural photochemical smog, which is formed by the action of sunlight on organic substances released from trees.

Carbon dioxide is emitted by humans, animals, combustion of fossil fuels, forest fires and decaying vegetation

Water condenses on to the sulphate particulates, and on to the dust and other particles that occur naturally in the atmosphere to form clouds

Oxygen is absorbed by humans, animals, combustion of fossil fuel, forest fires and decaying vegetation

Algae produce DMS, which escapes into the atmosphere forming sulphate particulates, which, in turn, help to form clouds

Evaporation of water from oceans to form clouds

Oxygen

Phytoplankton (algae)

Water runs off the land and returns to the oceans

CLOUDS

FROM WISPY EPHEMERAL FEATURES TO OMINOUS GREY
CUMULONIMBUS, CLOUDS DETERMINE OUR LOCAL
WEATHER AND THE EARTH'S RADIATION BALANCE.

Clouds are the greatest challenge for meteorologists in their efforts to explain the climate. They come in all shapes and sizes, and some only exist for a matter of minutes while others are effectively semi-permanent features. It is not just their variability that makes them so difficult to include in our thinking – their basic physical properties depend on a complex range of processes, which mean they exert a powerful influence on both incoming solar radiation and outgoing terrestrial radiation.

ALTERING THE RADIATION BALANCE

All clouds are efficient reflectors of sunlight. High thin clouds may reflect only one-fifth of incoming solar radiation back out to space, while low thick cumulus can reflect as much as three-quarters of it. Clouds are also very efficient absorbers of outgoing heat radiation from the Earth's surface and lower atmosphere. They act as a blanket keeping the heat in and radiate effectively at the temperature of their tops. This means that high clouds will radiate small amounts of energy because the temperature at their top is cold, while low warm clouds will radiate much more heat. However, low clouds have a much smaller impact on how much energy is emitted into space because they radiate from a level that is much closer to the ground.

Different types of cloud have different impacts on the temperature at the Earth's surface. Deep thick clouds, notably those found in the tropics, trap more radiation than they reflect sunlight, and therefore act as a warming influence on the climate. By way of contrast, low decks of stratus, found in depressions in mid-latitudes or over cold waters at low latitudes, are more effective reflectors than they are heat blankets, and so they cool the local climate. Globally, the combined effect of all clouds cools the temperature at the Earth's surface, but the impact of changes in cloudiness will depend on shifts in the proportion of different types of cloud.

HOW DO CLOUDS FORM?

An understanding of how clouds are formed and distributed is essential so that climatologists can accurately represent clouds in climate models for weather forecasting and predict how human activities may influence future climate conditions. At a simple level it all seems straightforward. Warm moist air rises and cools, and water vapour condenses out of the air to form tiny droplets at the level where the temperature is low enough for the air to become saturated: this is known as the dewpoint.

THE ROLE OF PARTICULATES

The process of condensation is, however, extremely complicated because it depends crucially on the presence of dust and aerosols in the atmosphere. Without a surface,

CLOUD FORMATION

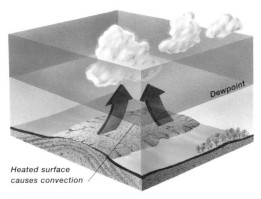

CONVECTION
When the ground is heated by sunlight, it causes convection in the air above it. The warm air rises and, when it reaches the dewpoint, condenses to form tiny water droplets, which are buoyed up by the rising air to form clouds.

Heated surface causes convection

Dewpoint

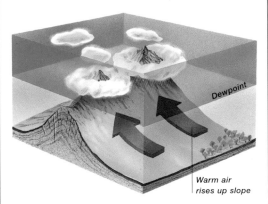

OROGRAPHIC CLOUDS
When air is forced to rise over hills and mountains, it cools and water vapour condenses to form clouds at high levels. These clouds often extend beyond the mountains themselves.

Dewpoint

Warm air rises up slope

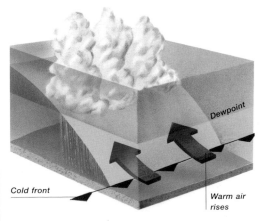

FRONTAL MOVEMENT
Air can be forced to rise by the movement of weather systems. Here advancing cold air causes warmer air to rise, forming clouds, which then precipitate through the colder air below.

Cold front

Dewpoint

Warm air rises

Left: High thin cirrus is often a sign of settled high-pressure conditions. This cirrus occurred over Dartmoor, England

Left: This towering cumulus shows that when the ground is well heated, warm air can bubble up to great heights to form distinctive cauliflower-shaped clouds, which often produce heavy showers.

condensation cannot occur so, if the air is completely clean, then droplets cannot form. In fact, this is never the case, but the shape, size and number of microscopic particles that are in the atmosphere determine the efficiency of the condensation process and how quickly the droplets grow within the cloud.

Over the land, there are many sources of the particulates in the air. A number of these come from natural materials, such as soil particles, aerosols formed from the vapour emitted by plants and sea salt that is blown inland. Particulates can also be formed as a by-product of human activities, such as smoke and the conversion of gas pollutants (for example, sulphur dioxide and unburnt hydrocarbons) into aerosols. At sea, the dominant source of particles is salt from the spray, which evaporates into the air, but the emission of sulphur-containing compounds by plankton is also significant.

Cloud formation is a complicated process because some particulates are much more efficient than others in creating either droplets or ice crystals. Some are better at the former while others are better at the latter. So, depending on what particulates are present in any parcel of air, clouds tend to form more readily in some cases than in others. To complicate matters still further, it is the type of particulate that determines whether the droplets or ice crystals grow rapidly and precipitate out as rain or snow, or stay

small and remain suspended in the cloud. The more particulates there are in the air, the more likely it is that a large number of droplets will remain suspended in the cloud and not fall to the ground.

The longer-lived the cloud, the more impact it will have on the radiative balance of the Earth. This means that fluctuations in the natural sources of particulates in the atmosphere, combined with the products of human activities, can have substantial impact on the number, extent, type and duration of clouds. All these variations can exert a significant influence on the climate.

THE IMPACT OF VOLCANOES

There are few natural phenomena as dramatic, and for humans as potentially devastating, as a volcanic eruption. Molten lava is the immediate danger but other emissions have longer-lasting effects.

Benjamin Franklin was the first to suggest that volcanoes could affect the climate. He proposed that the bitter winter of 1783–4 in northern Europe was due to the dust cloud from the eruption of Laki in Iceland in July 1783, which dimmed the Paris sky for months.

Dust veils and sulphur

Explosive volcanic eruptions can inject vast amounts of dust and sulphur dioxide into the atmosphere. Both have climatic repercussions. Sulphur dioxide has the greatest impact on the climate, as when propelled into the upper atmosphere it is converted into tiny sulphuric acid droplets. At altitudes of 15–30km (50,000–100,000ft), where there is no significant vertical air motion, these minute particles can remain suspended for several years and spread around the globe. In addition, solid particles form a dust veil in the upper atmosphere. Both the droplets and the dust veil absorb sunlight, therefore heating the stratosphere but cooling lower levels by reducing the amount of solar radiation reaching the Earth's surface.

Measuring volcanic eruptions

Analysis of how past volcanic activity affected the climate is problematic because worldwide historical records were not kept until the late 19th century and so it is hard to be certain of global changes. In the 20th century, there were no significant eruptions until 1963 when Agung in Bali erupted. But since 1980, three major but very different volcanic events have given climatologists the opportunity to discover much more.

The first of these was when Mount St Helens in the USA erupted in 1980. Contrary to popular belief, this had no notable climatic impact because it blew out sideways and did not inject much dust high into the stratosphere. More significantly, its plume was low in sulphur. Although it only had a limited effect in terms of cooling, it provided valuable insights into the role of sulphur.

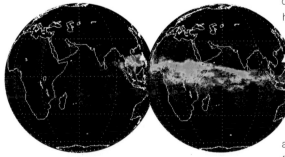

Above: Satellite images show how the dust cloud from Mount Pinatubo was swept around the world in just 10 days by winds in the stratosphere.

Left: The eruption of Mount Pinatubo in 1991 provided the clearest evidence yet of how volcanoes affect the climate.

The second eruption, of El Chichón in Mexico in 1982, though a fairly small volcano, was high in sulphur content. Satellite measurements of the stratospheric aerosol cloud, made up of tiny sulphuric acid droplets, confirmed the importance of sulphur. Finally, the most important eruption climatically was Mount Pinatubo in the Philippines in 1991, which injected around 20 billion kg (20 million tons) of sulphur compounds into the stratosphere.

Scientists used these eruptions to test climatic theories by making atmospheric measurements and using computer models of the climate. Accurate measurements of the temperature of the stratosphere by satellites showed how volcanic dust clouds warmed the upper atmosphere. The surface-temperature records revealed a cooling in the lower atmosphere, as predicted by climate computer models. These observations also confirmed that the effects of a single eruption can last for around two to three years.

Evidence from early volcanoes

Recent studies of Antarctic and Greenland ice cores and tree rings around the northern hemisphere have produced a more detailed picture of the link between volcanoes and the climate. These identify which eruptions had the greatest impact on the atmosphere, and the influence they had on weather patterns. In the northern hemisphere the greatest cooling effect is during the summer: winters are, if anything, warmed by the change in atmospheric circulation caused by eruptions.

In the last 600 years or so the coldest summer was in 1601, and appears to have been caused by a major eruption in Peru the year before. The so-called 'year without a summer' of 1816 was the second coldest on record and was probably due to the massive eruption of Tambora in Indonesia in 1815, which injected five to 10 times more material into the stratosphere than Pinatubo would later do. Exceptionally late frosts destroyed crops in New England, and French vineyards had their latest wine harvest for at least five centuries.

The short-lived impact of volcanoes suggests they can only trigger lasting change if they coincide with other factors that also stimulate cooling. For example, the largest eruption in the last million years – Toba in Sumatra, 73,000 years ago, which was at least five times the size of Tambora – coincided with a rapid cooling during the development of the last Ice age. This eruption could have produced sufficient cooling for perennial snow cover to form for a number of years in, say, northern Canada. At a time when the Earth was already slipping into a cold phase, this snow cover, with its additional cooling effect due to reflection of more sunlight into space, may have tipped the balance.

When Mount Pinatubo erupted in June 1991 after being dormant for more than 600 years, it took not only scientists by surprise but also its Filipino neighbours, whose life it touched in a dramatic way. The first inkling of Pinatubo's fury was a deep rumble, which sounded like thunder coming from inside the Earth. Then a dark angry mushroom-shaped cloud appeared in the sky and rained sand over the countryside and towns for miles around. After three days the rumbling became more furious and the sky turned black, as dark as night. By a cruel twist of fate, a typhoon struck at the same time, and the rain mixed with the volcanic ash and fell from the sky as mud. The sulphur in the air, the overwhelming stench and the huge fiery displays of lightning were an apocalyptic combination. Then the earthquakes began, bringing buildings crashing to the ground. Over 300 people were killed by the eruption and it destroyed the homes and livelihoods of thousands more.

THE SUNSPOT CYCLE?

LIFE ON EARTH IS SUSTAINED BY THE SUN AND SO ANY CHANGE IN ITS OUTPUT COULD HAVE AN EFFECT ON OUR CLIMATE.

Sunspots – darker patches found on the Sun's brilliant surface – are the best example of variation in solar activity, and have long fascinated meteorologists as a possible cause of climate change.

WHAT ARE SUNSPOTS?

Since the start of the 17th century, when Galileo first turned his telescope towards the Sun, we have known that it is blemished by sunspots, which erupt on its surface. These areas are cooler than the surrounding regions and can normally be observed at low latitudes between 30°N and S. They cross the face of the Sun in time with its rotation every 27 days. Each sunspot has two regions: a dark central umbra, which has a temperature of nearly 4000°C (7000°F) and a surrounding lighter penumbra, which has a temperature of nearly 5000°C (9000°F). The apparent darkness of the spots is purely a matter of contrast with the brightness of the Sun's brilliant surface temperature of nearly 6000°C (11,000°F).

Sunspots vary in their number, size and duration. There may be as many as 20 or 30 spots at any one time, and a single spot can range from 1000–200,000km (620–125,000 miles) in diameter with a life span ranging from a few hours to up to several months.

SUNSPOT CYCLES

In 1843, the German amateur astronomer Heinrich Schwabe made the discovery that the number of sunspots fluctuates in a regular, predictable manner. Both their number and their total surface area appear to follow a cycle that has an average length of 11.2 years. Additionally, sunspots exhibit another cycle of approximately 90–100 years, which is often termed the Gleissberg cycle.

The possibility that the energy output of the Sun could also vary in a periodic manner was, for a long time, the subject of great debate amongst scientists. Satellite measurements since 1980 have confirmed that the output of energy from the Sun does indeed rise and fall during both sunspot cycles – the more sunspots, the greater the amount of energy emitted.

DO SUNSPOTS AFFECT THE CLIMATE?

Reliable measurements of the changes in the Sun's energy output began in 1980 and so there is now data for nearly two 11-year cycles. It shows that the variation is only about 0.1 per cent. Theoretically, this is not sufficient to demonstrate a significant climatic impact. That said however, there does seem to be a link between global temperature and solar output as measured by the number of sunspots over longer timescales. Statistical analysis of these changes suggests that they could be responsible for up to half of the variation in the temperature of the northern hemisphere over the last four centuries.

Intense scientific debate continues as to whether solar activity can exert a greater influence on the climate. New computer models suggest that the absorption of solar radiation in the stratosphere could vary during the 11-year sunspot cycle and alter how much energy reaches weather systems at lower levels. This could have a significant impact on global circulation patterns, for instance by shifting the tracks of depressions from their usual course. Thus, despite the fact that there is only a small increase in the total energy emitted by the Sun when there are more sunspots, there seem to be significant changes in how much energy enters the Earth's lower atmosphere and where it does so – an amplification process could be at work.

CAUSES OF THE AMPLIFICATION PROCESS

The amplification process can be explained in two ways. Firstly, much of the change in solar output during sunspot cycles is concentrated in the ultraviolet (UV) part of the spectrum – energy which is absorbed high in the atmosphere by both oxygen and ozone molecules. This has a disproportionate effect on the stratosphere which may lead to changes in the behaviour of the jet stream in the upper levels of the troposphere. Changes in the amount of UV radiation reaching the lower atmosphere can also affect cloud formation. At times of high solar activity more UV radiation reaches the troposphere, and this may lead to an increase in the concentration of cloud condensation nuclei, making our skies cloudier. This process may reinforce any fluctuations in global circulation patterns caused by varying solar activity.

Secondly, magnetic fields associated with the sunspot cycle could be the key to the amplification process because the polarity of sunspots alternates between positive and negative in successive 11-year periods. Named after the astronomer Hale, this 22-year cycle influences the strength of the magnetic field between planets, which controls the interaction between the solar wind and the Earth's magnetosphere. This combines with the Earth's magnetic field to affect the amount of cosmic rays (energetic particles from the Sun and elsewhere in the universe) so that more particles are funnelled down into the atmosphere when sunspot numbers are high. This influence of the magnetic field on the climate may explain a widespread but faint 20-year cycle in many weather records.

RECORDS OF SUNSPOTS

Astronomical observations since the end of the 17th century show that there has been a clear cycle in the number of sunspots, each with an average period of about 11 years. Superimposed on the shorter fluctuations in sunspot numbers is a curve showing an average over about 20 years, which enables us to consider variations that occurred for longer periods of time. From this we can see that overall sunspot activity was low at the beginning of the record, in the early 19th century and at the beginning of the 20th century.

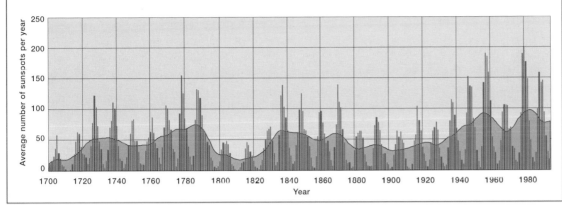

PAST FLUCTUATIONS

Long-term variations of solar activity are revealed in the amount of radiocarbon found in tree rings. The strength of the solar magnetic field affects the production of radiocarbon in the Earth's atmosphere because it controls the influx of cosmic rays. Tree-ring records going back some 9000 years provide evidence of periodic fluctuations in solar activity. Studies carried out on data from the White Mountains in California and from fossilized oaks in Europe suggest that there are five main longer cycles in solar activity with periods of about 2300, 500, 355, 204 and 154 years. The existence of the 200-year cycle is intriguing because over the last 1000 years the troughs in radiocarbon production indicate lulls in sunspot numbers at around AD1280, AD1480 and AD1680 and, to a lesser extent, at the end of the 19th century. These periods of low solar activity seem to coincide with cooler periods in the northern hemisphere, notably the end of the 17th century.

GALILEO GALILEI (1564–1642)
This famous Italian astronomer, philosopher and scientist made a number of significant contributions to climate studies, the most important of which was his invention of the thermometer in c.1600. (His pupil, Evangelista Torricelli, invented the barometer in 1643.) He also rediscovered sunspots – previously observed by Chinese astronomers 2000 years ago – and in a famous treatise in 1613 described how they moved across the face of the Sun as it rotated. He is perhaps best known for his refinement of the recently discovered refracting telescope, and his many astronomical discoveries, which led him to conclude that the Earth orbited the Sun and thus brought him into conflict with the Church.

Left: A photograph of the Sun in visible light, showing the cooler area of the sunspot as a dark blemish.

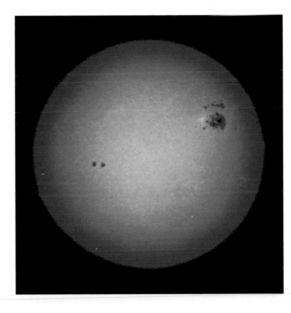

Left: A false-colour image of a cluster of sunspots, showing the Sun's normal surface as yellow and the sunspots themselves as purple and black.

Far left: This image of the Sun was taken by the SOHO spacecraft, which is positioned some 1,500,000km (930,000 miles) from the Earth at the point where the gravity pulls of the Sun and Earth are equal. Thus, SOHO orbits the Sun in the same way that the Earth does.

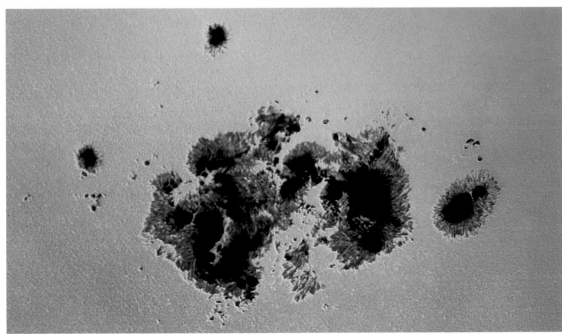

ATMOSPHERIC ELECTRICITY

AT THIS MOMENT THERE ARE APPROXIMATELY 2000
THUNDERSTORMS RUMBLING AROUND THE WORLD,
CAUSING 100 LIGHTNING FLASHES EVERY SECOND.

BENJAMIN FRANKLIN (1706–90)
Benjamin Franklin, one of the
founding fathers of the USA, was
a formidable polymath who made
many important contributions to
early meteorology. Apart from his
famous experiment in 1752 to
demonstrate the electrical nature
of lightning and his invention of
the lightning conductor, he
studied the nature of clouds and
rainfall. He also made pioneering
observations on the movement of
weather systems. He was intrigued
by the fact that he was unable to
observe an eclipse of the Moon in
Philadelphia on 21 October 1743
because of stormy weather,
whereas his brother in Boston had
a clear view because the clouds
did not arrive until two hours later.
He collected newspaper reports
and, from these, established the
rate at which the storm had
travelled up the east coast.

The role that atmospheric electricity plays in the Earth's climate system remains one of the greatest unsolved mysteries of meteorology. At a microscopic level, it is clearly central to the properties of thunderstorms, as is apparent in the spectacular form of lightning strikes. At a global level its impact remains a matter for speculation, although the incidence of lightning may be interpreted as a sensitive indicator of climate change.

INCREASING AWARENESS OF LIGHTNING

Benjamin Franklin proved that lightning was an electrical discharge and measured the sign of the cloud charge that produced it more than 200 years ago. However, progress on lightning research was slow until the 20th century. In the 1930s, it was aimed principally at protecting electric power-supply systems from lightning strikes. It was not until the 1960s that the vulnerability of modern electronics became clear, with a Boeing 707 crash in Maryland in 1963, which killed all the passengers. In 1969 the Apollo 12 spacecraft initiated lightning during its take-off procedure, but fortunately it did not strike or damage the launch vehicle.

HOW MUCH ELECTRICITY IS GENERATED?

Lightning generates huge amounts of electricity: it can involve voltage differences as great as 10 million volts, leading to high currents of between 100 and 1000 amps or, in certain cases, of as much as several hundred thousand amps. Such intense currents can generate temperatures in the lightning bolt of as high as 20,000–30,000°C (36,000–54,000°F) – several times hotter than the Sun.

Over half of lightning strikes occur within clouds but strikes to the ground are more visually sensational and are responsible for the most death and destruction. In the USA the number of fatalities has fallen from around 400 a year in the early part of this century to 50 to 60 in recent years because of the decline in rural population and the number of people working on the land. Nevertheless, the number is still comparable with the annual toll due to tornadoes and considerably higher than the death toll from hurricanes.

Right: Intracloud lightning over Phoenix, Arizona, USA. These cloud-to-cloud discharges are the most common form of lightning.

HOW DOES LIGHTNING HAPPEN?

Thunderclouds generate electricity from the separation of positive and negative charges by ice crystals and water droplets. These gain their charges in a number of ways. Firstly, when a water droplet freezes rapidly, it gains a positively-charged outer shell. If this shatters, the light ice crystals with positive charges are swept upwards while the heavy inner core with a negative charge falls to the ground (a). Secondly, a small ice crystal acquires a positive charge when it collides with a large ice pellet (b). In this way, light ice crystals carry positive charges to the top of the cloud (c), and heavy ice pellets and water droplets carry negative charges to the bottom of the cloud (d). These negative charges induce positive charges in the ground. At its simplest level, when the charge builds up to a sufficient level it can break down the resistance of the air and produce a bolt of lightning.

CLOUD-TO-GROUND LIGHTNING STRIKE

1 Streamers of negative charge are attracted by the positively-charged ground. This in turn encourages positive streamers from the ground.
2 When the two streamers meet, a powerful stroke of lightning occurs upwards from the ground.
3 The upward positive current sets off a return stroke from the cloud back down to the ground. This process is repeated, which causes the characteristic flickering effect of lightning.

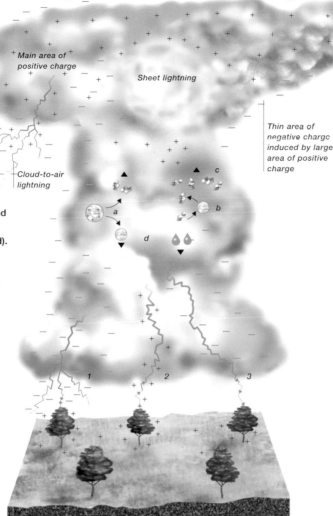

Above: An intense bolt of lightning like this is able to generate temperatures as high as 30,000°C (54,000°F).

MONITORING LIGHTNING STRIKES

Satellite measurements of flash rates show that the majority of storms occur over land in the tropics because the land is warm and stimulates deep convection, which generates strong updrafts of air that produce the right conditions for charge separation. Southeast Asia and Australia, Africa and South America are the three major zones of deep electrically active convection.

Ground-based lightning-location networks can measure the frequency and position of thunderstorms at distances of more than 100km (60 miles) by using magnetic direction finders and time-of-arrival receivers to detect the radio signals that are emitted by lightning. These networks are used for scientific studies of lightning physics and the properties of storms, for engineering studies associated with electric power-transmission systems and for warnings of forest fires during periods of drought.

LIGHTNING AND CLIMATE CHANGE

The amount of lightning activity is closely associated with the temperature of the atmosphere in the tropics. As the temperature rises above 25°C (77°F), more deep convection with strong upward currents of air occurs, and this greatly increases the chances of lightning strikes occurring. Thus, any change in the amount of lightning throughout the tropics is a sensitive indicator of temperature variations. Very low-frequency radio signals that are emitted from lightning flashes can travel all round the world, and so the size of these signals is a measure of the total amount of global thunder activity at any one time. Changes in these signals over time will, therefore, indicate shifts in the temperature of the tropics. Furthermore, thunderstorm activity is linked to other important climatic factors, such as the intensity of the intertropical convergence zone (ITCZ), and the formation of cold clouds at high levels in the atmosphere, and for this reason may prove to be a valuable method of detecting certain features of global warming.

Another way in which lightning plays a part in climate change is through its links with solar activity. The changes in the flux of cosmic rays, caused by rising and falling levels of solar activity, could affect the Earth's electric field and alter the incidence of thunderstorms. This could magnify the impact of solar activity on the climate. Moreover, lightning is a major natural source of oxides of nitrogen and ozone, and these gases make an important contribution to the greenhouse effect. Therefore, if global warming leads to an increase in the number of thunderstorms occurring around the world, this could serve to reinforce the warming trend – a positive-feedback mechanism.

THE GAIA PRINCIPLE

IT IS WELL KNOWN THAT LIFE ADAPTS TO THE ENVIRONMENT, BUT COULD LIFE ALSO REGULATE GLOBAL CONDITIONS TO KEEP IT HABITABLE?

The intriguing but controversial Gaia hypothesis was first proposed by James Lovelock and named after the 'mother Earth' goddess of the ancient Greeks. It aims to explain why, unlike other planets in the solar system, the Earth's atmospheric composition and history cannot be described in terms of physics and chemistry alone, but reflect the strong influence of biology.

SURVIVING AGAINST ALL ODDS

Life on Earth has survived for nearly four billion years, in spite of huge changes in the physical conditions affecting the planet, such as the rising output of the Sun, continental drift and bombardment by asteroids. Lovelock argued that this durability could only be explained by treating life and the global environment as two parts of a single system. In effect, micro-organisms, plants and animals behave in such a way that the Earth's environment becomes adjusted to states optimal to the maintenance of their existence. This is not a conscious act on the part of the biosphere but adjustments arise from the process of natural selection.

A good example of the type of process that might contribute to the stability of the global ecosystem is the production of dimethylsulphide (DMS) by phytoplankton (algae) in the oceans. When DMS is released into the atmosphere, it is converted into sulphate particulates, which aid cloud formation and hence influence the climate. So, for instance, if the world were to get warmer, sea-surface temperatures and carbon dioxide levels would rise, more algae would be produced and as a result more DMS would be released into the atmosphere. This, in turn, would result in more clouds, which would have a cooling effect on the climate. In effect, the response of the algae to these changes is to act as a climatic 'thermostat'.

CONTROVERSY

The Gaia hypothesis has been the subject of controversy since publication in 1972, though there is no doubt that the presence of life on Earth has had a profound effect on the climate. The fact that the output of the Sun has risen by about 20 per cent over the last four billion years should have led to substantial warming, but, in practice, the temperature has remained remarkably constant. This stability has been maintained largely by a decline in the levels of carbon dioxide in the atmosphere from levels several hundred times higher than they are today.

The essence of the controversy is the extent to which life is in the driving seat in these changes, as opposed to volcanoes and the continual process of weathering of rocks. The danger of the hypothesis is to regard Gaia as some form of huge organism controlling all aspects of the climate. Opponents of this interpretation argue that it requires all life forms to have evolved altruistically, for the

JAMES LOVELOCK (1919–)
A British chemist, James Lovelock is best known for his Gaia hypothesis, which generated a huge storm of controversy amongst scientists. One of his less controversial contributions was the electron capture detector, which he invented in 1958. This sensitive device is able to measure chemical compounds in the Earth and atmosphere with an extremely high degree of accuracy. Lovelock carried out his own experiments to detect the levels of CFCs in the atmosphere and his observations eventually led to the first suspicions that CFCs were damaging stratospheric ozone. He has also worked with NASA investigating the possibility of whether life exists on Mars.

Above: The biosphere from space, showing productivity in the oceans (dark blue for the lowest levels to greens, orange and red for the highest) and vegetation on land (yellow and orange for the sparsest to blue-green for the most dense).

Right: The Earth as seen from the Apollo 17 spacecraft in 1972.

good of the whole. Lovelock's simple models conceive of evolutionary processes working to stabilize the climate. However, in the real world the complexity of life makes it hard to see how organisms of such bewildering diversity could have co-operated for all their mutual benefit.

At a more specific level, experiments to prove theories such as the DMS hypothesis have been inconclusive. For instance, measurements in the tropical Pacific have shown little evidence of the amount of DMS in the atmosphere being associated with annual changes in sea-surface temperatures. Similarly, while computer models have been developed to explore the validity of the theory, there is always the possibility that scientists select the right circumstances to reinforce the case for the hypothesis.

POSITIVE AND NEGATIVE FEEDBACK

Lovelock proposed an imaginary world (Daisyworld) in which the only form of life is black and white daisies and the only thing that affects them is temperature. They grow best at 20°C (68°F), will not grow below 5°C (41°F) and will die at over 40°C (104°F). Their colour can influence the temperature: white daisies reflect more sunlight and cool the planet, while black ones absorb more and warm it up.

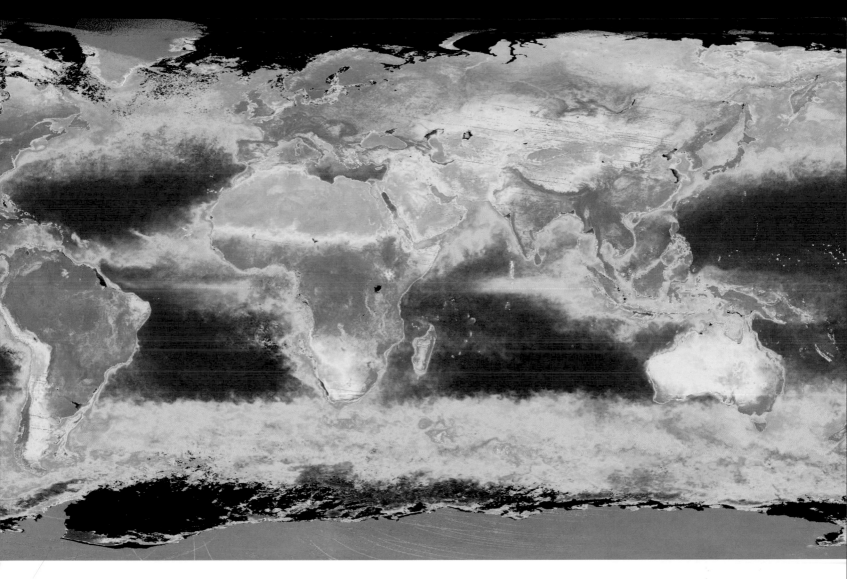

sunlight, not overheating like the black daisies, and cooling their surroundings at the same time. So they will compete successfully and start to thrive, preventing the planet from overheating: a negative feedback process.

As Daisysun warms up, the balance between black and white daisies will shift in favour of white daisies. But, eventually, it will get too warm for even white daisies to keep the temperature below 40°C (104°F) and they will die. Until this happens the fluctuating population of daisies holds the planet's temperature within a liveable range.

This simple model created by Lovelock shows how the Gaia principle operates and also contains a fundamental climatic truth: negative feedback is a stabilizing influence because it counteracts the effect of change, whereas positive feedback is unstable because it contributes to the runaway response.

HOW WE LOOK AT THE CLIMATE

In one important respect the objections to Lovelock's theories are irrelevant. In considering the global climate it is crucial to be aware of the intricate links between all the components of the system and the Gaia concept of the biosphere provides an intriguing way of looking at this complexity. Leaving aside the controversial aspects of the theory, it helps us to address the overall response of the climate system and the biosphere to change. So, however we look at the history of the Earth's climate, it is helpful to consider its overall response to all forms of change.

If, like our Sun, Daisyworld's Sun (Daisysun) gets hotter as it ages, then the black daisies will have an advantage early in the planet's life, multiplying and extending their range to the cooler parts of the planet. But, as they become the dominant population, they will heat the planet up so that the equatorial regions will become too warm for plants that absorb sunlight efficiently: a runaway process that is termed positive feedback. The white daisies, however, will be able to exploit the warming process by reflecting more

Above: The white daisies reflect more sunlight than the surrounding darker vegetation and indicate how the imaginary Daisyworld might function.

CLIMATE RECORDS

CURRENT MEASUREMENT SYSTEMS

THE KEY TO UNDERSTANDING AND PREDICTING THE
WEATHER LIES IN OUR ABILITY TO MAKE RELIABLE
AND DETAILED RECORDS ALL OVER THE GLOBE.

Our detailed knowledge of the climate is built on the painstaking accumulation of instrumental recordings, which have been made over many years. We also need to measure current conditions correctly to be able to produce accurate weather forecasts.

MODERN TEMPERATURE MEASUREMENTS

Modern instruments are capable of making very accurate observations of the climate. The challenge is to ensure they measure the same factor under the same conditions, wherever and whenever they are used. In the case of temperature measurement, for instance, a properly calibrated thermometer making ground-level observations must record the shade air temperature, not the capacity of the thermometer to absorb sunlight or the potential of the ground to heat up and so influence what is observed. Thermometers must therefore be mounted in a well-ventilated, louvred, white shelter, which prevents either direct sunlight or terrestrial radiation reaching the instruments. The standard design for such an installation is known as a Stevenson screen (see box, far right).

PAST TEMPERATURE MEASUREMENTS

Although measurements from modern instruments are reliable, earlier observations were often made in different conditions and so must be viewed with caution when comparing them with current figures. Additionally, changes in land use over long periods can exert significant influences on local temperatures. In particular, many early observations were made in the vicinity of towns, which had developed an urban heat island. Although this warming is real, it must not be confused with wider changes in the climate. Similarly, the movement of an observation site from a city centre to the local airport could lead to a significant shift, which must also be taken into consideration.

Measurements of sea-surface temperatures provide a different set of challenges. In the 19th century many ships kept records of the temperature of water collected in buckets over the side of the ship, but meteorologists need to know whether the samples were collected in wooden or uninsulated canvas buckets because the latter cooled more quickly. Other variables have to be identified and allowed for in order to get an accurate estimate of changes over the last 150 years. For example, during World War II there was an undocumented shift in practice, from using canvas buckets to measuring the temperature of the water taken in to cool the ship's engine. This was because it was too dangerous to use lights to make measurements over the side at night during wartime.

Observations made in the 19th century are more of a challenge, although there are some recordings of a reasonable standard available, and, when these are cross-referred with overlapping records, reliable figures can be obtained. Adjustments are then made: for example, to compensate for measurements taken at different times of day. These records provide temperature measurements for Europe and eastern North America. The longest is one for rural sites in central England prepared by Professor Manley, who brought together all the records that had been accumulated by a bewilderingly diverse array of amateur observers before the days of official meteorology.

Right: A digitally enhanced image showing an aircraft that is used to measure lightning. This type of aircraft has to be strengthened and shielded in order to carry out the research.

12 Outgoing energy

Above: Countdown to the release of a weather balloon near Antarctica. This is part of a routine meteorological observation used to measure upper-atmosphere winds.

Observations from before 1760 pose still greater problems because they are even less reliable. Some of the best-kept records depended on having thermometers exposed in well-ventilated north-facing fireless rooms.

RAINFALL MEASUREMENTS

Rainfall measurements on land have been made for at least as long as temperature observations. Here again there are problems with the reliability of early observations, many of which probably understate rainfall amounts because of the siting of gauges: any interference of swirling wind near the gauge is capable of distorting the amount of rain caught. Wind has an even greater distorting influence on snowfall and early measuring systems failed to catch much precipitation. These problems are compounded not only by the growth of vegetation and/or buildings around the site, but also by any change in the design of the instruments, which can alter the reliability of the observations.

While the most populous areas of the world have built up reasonable rainfall records, elsewhere there are huge gaps. So it is not possible to say precisely how much rain falls around the world or whether this amount is rising or falling with global warming. On a regional basis there is some evidence of precipitation rising in mid-latitudes as a result of stronger westerly circulation, while in the subtropics of the northern hemisphere there has been a pronounced drop in recent decades, attributed almost entirely to the drought in sub-Saharan areas of Africa.

OTHER MEASUREMENTS

Observations of other meteorological factors, such as atmospheric pressure, wind speed, cloud cover and sunshine, present the same set of challenges when trying to ensure consistency from one place to another and over

a period of time. Furthermore, lengthy records are not available in many places and so it is difficult to establish whether there have really been significant changes over the course of the centuries.

The development of modern technology complicates matters still further. Weather conditions throughout the depth of the atmosphere (surface to 30km/100,000ft, above ground level) are now regularly measured using balloon-borne instrument packages and radio transmission (radiosondes). Although global coverage is spotty, these measuring systems have provided invaluable information for scientists and forecasters. However, the equipment has undergone various design changes over the last 30–40 years, which have made the data collected inconsistent. This means that it is difficult to identify precisely long-term changes in global conditions in the upper atmosphere.

A GLOBAL NETWORK

World Weather Watch is a measuring system that uses the facilities of the national weather services of over 170 of the countries belonging to the World Meteorological Organization (WMO). This involves not only the network of some 12,000 land stations and 700 radiosonde-launching sites, but also measuring equipment on ships, including the WMO Voluntary Observing Fleet's 7300 merchant vessels, automatic buoys, commercial aircraft and weather satellites. All these data are transmitted via teletype and radio links to regional or national centres, and are then fed into the Global Telecommunications System connecting the three World Meteorological Centres – one in Melbourne, one in Moscow and one in Washington DC – which assemble data and produce synoptic weather maps, similar to those we are familiar with on television, every six hours.

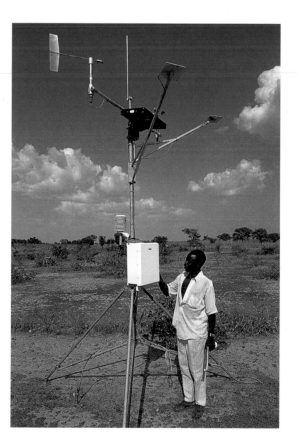

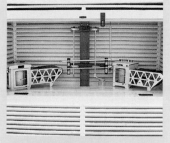

STEVENSON SCREEN
The standard shelter for meteorological instruments, the Stevenson screen should be positioned 1.2m (4ft) above short grass and at least 10m (33ft) from any buildings. It normally holds maximum and minimum thermometers, an aneroid barometer, and wet- and dry-bulb thermometers (for measuring relative humidity). Larger shelters may contain a thermograph and hydrograph, which measure fluctuating temperatures and precipitation. The screen has a double roof, and a slatted base and sides to increase ventilation. It is white so that thermometers within can give a close approximation to the true air temperature, undisturbed by the effects of solar or terrestrial radiation.

Left: Monitoring climatic change in Silmiougou Village in Burkina Faso.

WEATHER SATELLITES

SATELLITES PROVIDE US WITH A WEALTH OF DATA
THAT HELPS US BOTH TO PREDICT THE WEATHER AND
TO MONITOR ANY CLIMATE CHANGE.

Above: The European Remote-Sensing satellite over the Netherlands coast. This satellite takes measurements in the infrared, microwave and radio wavelengths and is used to study ocean currents and sea-surface temperatures. It is also used to monitor the extent of vegetation cover and polar ice caps, and to detect oil spills.

Weather satellites are an integral part of our lives. The images they produce are used daily throughout the world in weather forecasts on television, their measurements provide new insights into the climate and their global coverage help us understand climate change.

THE EARLY DAYS

The first of a generation of weather satellites, the hat-box-sized American TIROS (Television, Infra Red Observational Satellite) spacecraft was launched in April 1960. Its grainy pictures were a far cry from today's daily display of multi-coloured three-dimensional moving images shown to millions of viewers. Nevertheless, within five years of the first TIROS, polar-orbiting satellites were sufficiently reliable to form a regular complete picture of almost the whole of the global atmosphere.

In 1966 there was another important development. Professor Vern Suomi, of the University of Wisconsin, conceived a system to exploit for meteorological purposes the spinning geostationary-satellite technology used for communications systems. He designed a television camera that was capable of scanning the face of the Earth. It was installed in the first experimental geostationary satellite, which was launched in December 1966. At an altitude of 35,800km (22,300 miles) above the equator, this system provided images every half hour, whereas the polar-orbiting satellites could only provide images of each part of the globe every 12 hours.

The value of weather satellites was demonstrated in dramatic fashion in 1969. The US National Weather Service was using satellite images to predict where and when hurricanes would strike land. When Camille, the most intense hurricane ever to hit the US mainland, reached the Louisiana–Mississippi coast the loss of life was relatively small because most of the resident population had heeded the warnings and taken shelter inland.

DIFFERENT MEASUREMENT TECHNIQUES

Big advances followed with the development of new measuring techniques. The early instruments had relied on measuring sunlight that was reflected from the Earth, which limited observations to the daytime and made it impossible to see below the clouds or to view the polar regions in winter. The big step forward was the discovery that the Earth's surface, the atmosphere and the clouds all emit heat radiation. Satellites are now able to detect heat and can make observations in various parts of the electromagnetic spectrum. This is known as temperature sounding. It enables observations to be made at night and also makes it possible to measure the temperature of the Earth's surface and the tops of clouds, and to make soundings at some levels in the cloud-free atmosphere.

Observations are usually made in the infrared range where most heat radiation is emitted, but even at this wavelength the clouds are opaque and obscure the ground. However, super-sensitive instruments can detect the tiny amounts of microwave energy emitted by the Earth and atmosphere, making it possible to 'see' through the clouds. Upwelling radiation is measured with radiometers, which scan the Earth below the satellite, sweeping back and forth and building up a picture of the amount of energy radiated by each part of the planet. The amount of energy emitted is related to the temperature of the surface or atmosphere below the satellite and so the measurements can be converted into local temperature figures. Different temperatures are represented on a grey scale or as different colours to show the conditions in the satellite's field of view.

The amount of detail in images varies. The satellites of the USA's National Oceanic and Atmospheric Administration (NOAA), which orbit at an altitude of about 900km (560 miles), measure the energy from 1km (0.6 mile) blocks and so produce high-resolution pictures. Geostationary satellites have a resolution of 5–10km (3–6 miles), while temperature soundings of the atmosphere and microwave observations are restricted to a few tens of kilometres.

HOW ARE SATELLITE DATA USED?

All of this information is used by forecasters to improve their weather predictions. The images of cloud cover enable them to check how accurately the computer models are reproducing the current state of the atmosphere or whether there are any tell-tale signs of rapidly emerging developments that are not picked up in the standard computer-generated weather forecasts.

Weather satellites can also be used to measure the extent of vegetation around the world through the seasons. This is possible because vegetation and open ground emit different proportions of visible light and infrared radiation. These measurements can monitor the effects of drought in different parts of the world, provide indirect measurements of rainfall and offer insights into how these changes are linked to other climatic factors such as sea-surface temperatures across the oceans.

THE LATEST DEVELOPMENTS

Experimental satellites have extended observations still further. Some satellites, such as the Upper Atmosphere Research Satellite (UARS), measure ultraviolet sunlight and can detect human pollutants and monitor changes in ozone levels in the upper atmosphere. Radar equipment can measure ocean waves and wind speeds, as well as observing the strength of ocean currents and the effects of the tides. They can also measure changes in the height of the polar ice sheets and detect global rises in sea level.

Satellites can now detect the extent of snow and polar pack-ice throughout the year, by measuring microwave emissions. This monitoring programme has provided the first reliable figures showing how the area of these climatically sensitive features has changed since the 1970s.

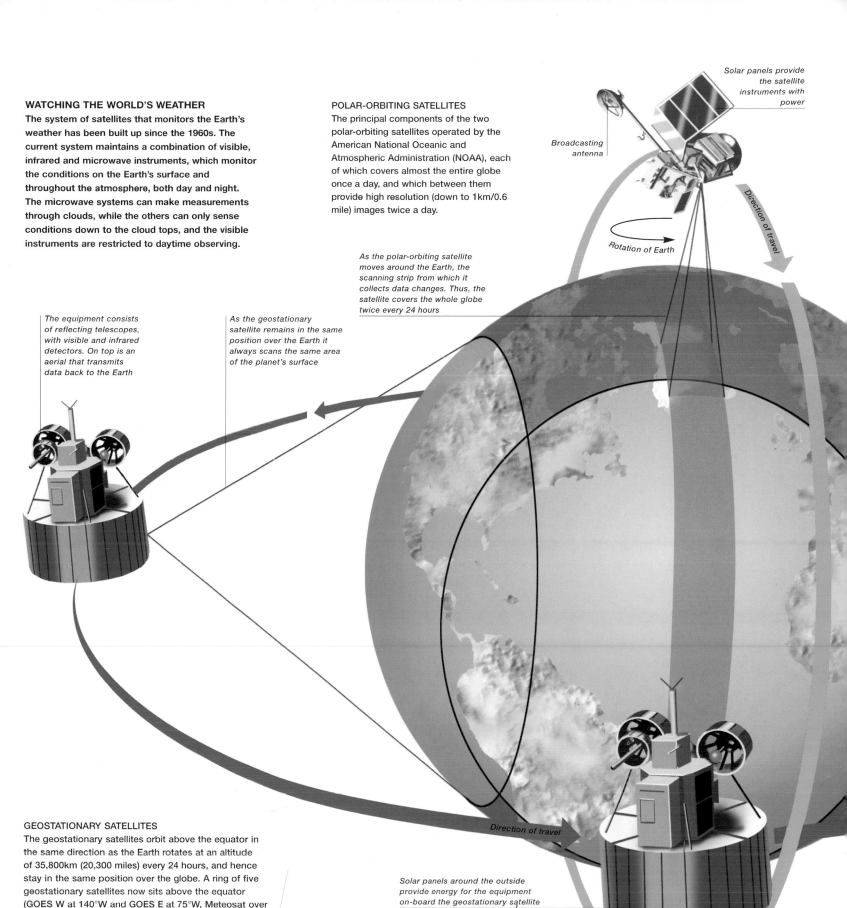

WATCHING THE WORLD'S WEATHER

The system of satellites that monitors the Earth's weather has been built up since the 1960s. The current system maintains a combination of visible, infrared and microwave instruments, which monitor the conditions on the Earth's surface and throughout the atmosphere, both day and night. The microwave systems can make measurements through clouds, while the others can only sense conditions down to the cloud tops, and the visible instruments are restricted to daytime observing.

POLAR-ORBITING SATELLITES

The principal components of the two polar-orbiting satellites operated by the American National Oceanic and Atmospheric Administration (NOAA), each of which covers almost the entire globe once a day, and which between them provide high resolution (down to 1km/0.6 mile) images twice a day.

Solar panels provide the satellite instruments with power

Broadcasting antenna

Direction of travel

Rotation of Earth

As the polar-orbiting satellite moves around the Earth, the scanning strip from which it collects data changes. Thus, the satellite covers the whole globe twice every 24 hours

The equipment consists of reflecting telescopes, with visible and infrared detectors. On top is an aerial that transmits data back to the Earth

As the geostationary satellite remains in the same position over the Earth it always scans the same area of the planet's surface

Direction of travel

GEOSTATIONARY SATELLITES

The geostationary satellites orbit above the equator in the same direction as the Earth rotates at an altitude of 35,800km (20,300 miles) every 24 hours, and hence stay in the same position over the globe. A ring of five geostationary satellites now sits above the equator (GOES W at 140°W and GOES E at 75°W, Meteosat over the Greenwich Meridian, plus the Indian satellite [INSAT] at 70°E and the Japanese satellite [GMS] at 140°E), providing half-hourly images of the globe between latitudes 65°N and S.

Solar panels around the outside provide energy for the equipment on-board the geostationary satellite

WEATHER FORECASTING

WEATHER FORECASTS HAVE BECOME SUCH AN
INTEGRAL PART OF EVERYDAY LIFE THAT WE RARELY
STOP TO CONSIDER HOW THEY ARE PUT TOGETHER.

Below: A satellite image of Hurricane Fran taken by NOAA's GOES-8 satellite in 1996, seven hours before it struck the coast of the USA at Cape Fear in North Carolina.

With entire television programmes and even entire channels dedicated to the subject, there seems to be no end to our insatiable appetite for weather and weather forecasting. Modern numerical weather prediction depends on supercomputers, which model global weather systems and at the moment are able to produce meaningful forecasts for up to about a week ahead.

conditions. These come from a variety of sources including ground stations, aircraft, radiosonde balloons, ocean buoys, and orbiting and geostationary weather satellites. All this information is then fed into the GCM. This process is not straightforward because the vast amounts of data will have been collected at slightly different times and do not coincide exactly with every point on the grid. As the variety of input data rises, calculating these initial conditions is becoming the most demanding part of the entire operation.

Next, the model calculates how the atmosphere will behave, using the laws of physics and taking into account other factors. These include the rotation, geography and

GATHERING THE DATA
Weather forecasts are achieved by using Global Computer Models (GCMs), which simulate and predict weather systems. They treat the global atmosphere as a multi-layered grid made up of four million points. More detailed models made up of more points are used for limited areas of the world in order to simulate the atmosphere and to produce local forecasts of rainfall. The first stage of the process is to make full use of all the available observations of current

topography of the Earth; the incoming solar radiation and its daily and seasonal variations; the radiative and absorptive properties of the land surface according to the nature of the soil, vegetation, snow and ice cover, and the surface temperature of the oceans based on current measurements, often relying on satellite measurements.

These calculations of the GCMs also check the accuracy of the input data by comparing new measurements and current model figures. For instance, satellite measurements of temperature sometimes blur out important details of the atmosphere, such as fronts, and so it is important to detect when the model and reality are drifting apart. This is done by comparing what the GCM thinks the satellite should see and what is actually measured, and, when there is a major difference between the two figures, it nudges the temperature values for various levels in the model back into line.

THE FORECASTS

Models can represent the state of the atmosphere in steps of a few minutes. How these changes build up over the days ahead is the basis of the standard forecast. A set of forecasts up to 10 days ahead involves some several tens of trillions of calculations and shows how different weather features such as temperature and atmospheric pressure at different levels, wind speed, cloud cover and precipitation will develop.

With the current number of observation points in the northern hemisphere, useful forecasts can be produced for about seven days ahead in the northern hemisphere and about five days in the southern hemisphere. At present, it is impossible to make accurate day-to-day forecasts more than about two weeks in advance because any errors in the measurement of the initial state of the atmosphere double in the computational process every two or three days. The more accurate the starting point, the more accurate the final forecast will be.

The computer data are then used to put together a series of maps that indicate various climatic features including atmospheric pressure, temperature and wind speed and direction. These frequently appear on television, in newspapers and on the Internet, and are used by presenters in the media to explain what the weather is going to be like in the next few days.

More specialized predictions can be extracted from the computer products by trained weather forecasters. This is particularly common in industries whose activities are sensitive to specific features of the weather. For example, airlines purchase predictions of upper-atmosphere winds, while shipping companies and offshore industries require forecasts of both wave heights and the underlying swell from distant weather systems (the latter is capable of travelling thousands of kilometres in a few days without calming down significantly).

IMPROVING FORECASTS

Better weather forecasts depend on improved measurements of current conditions and a more accurate physical representation of how the atmosphere interacts with the Earth's surface. Only by finding out exactly how much energy and moisture are exchanged through radiative and convective processes is it possible to produce realistic models. Therefore, studies based on the effects of soil moisture, snow cover, mountain roughness and the interplay of wind and waves across the oceans are essential to making progress in this field. There has been a large-scale international effort in recent years to examine these surface processes at work – experiments have been carried out over the prairies of the USA, the forests of Canada, and over the Pyrenees between France and Spain. At the same

time, improvements in weather satellites have given weather forecasters access to much better observations of wind and wave movements.

These measurements confirm that the amount of energy exchanged is critically dependent on the surface conditions. Meteorologists need to know how much sunlight is absorbed into the Earth's surface, how much is re-radiated back out to space and how much remains in the lower atmosphere, converting moisture into water vapour and thus forming clouds. Saturated ground can exert a major influence on the properties of the lower atmosphere in the summer. For example, unprecedented rainfall over the Midwest of the USA during the summer months of 1993 led to the record-breaking floods that occurred along the Mississippi, which were not accurately predicted. Meteorologists then went back and incorporated the soil-moisture effects into the computer models and the forecast of rainfall improved spectacularly.

An improved understanding of how mountains interact with different weather systems will enable forecasters to produce better predictions of where the heaviest precipitation is likely to occur, both in terms of how eddies form around the mountains and how more rain will fall on the windward slopes rather than around the peaks. This has considerable potential for providing improved warnings of flash flooding.

A MULTITUDE OF FORECASTS

The quality of weather forecasts is sensitive to how unstable the atmosphere is when the observations are made. A new technique, known as ensemble forecasting, can now determine how unstable the atmosphere is at any one time. This is done by examining how the predictions respond to slightly different starting conditions. If, with a subtle set of variations, an ensemble of forecasts looks remarkably similar up to about 10 days ahead, then there is a good chance they are on the right track. If, however, the forecasts diverge significantly, then the atmosphere is in a less predictable mood. This forecasting technique is now an integral part of the weather forecaster's armoury.

Above: A view of the operations in the air-traffic control tower at Memphis Airport, USA. Overlaid on this is a frame from the National Lightning Detection Network computer, showing the distribution of lightning strikes (green dots) across the USA. Air-traffic controllers use this information in planning the most efficient routes for their aircraft.

HISTORICAL RECORDS

HISTORICAL RECORDS ARE IMPORTANT TO OUR
UNDERSTANDING OF PAST CLIMATIC CHANGE – THEY
ALSO ADD COLOUR AND CONTEXT TO PROXY DATA.

Historical records, such as personal diaries and agricultural records of the condition and price of crops, are useful sources of information about past weather, especially where instrumental observations do not exist.

WHEAT PRICES

In Europe and China detailed reports of many annual events going as far back as the Middle Ages provide valuable insights into the weather conditions at the time. This is particularly useful where historical records overlap with instrumental observations and so can be checked against reliable measurements. Their frequent dependence on agricultural activity is, however, a limitation. For example, the price of wheat is a good indicator of the abundance of the harvest because bumper years led to low prices and a dearth produced high prices. But, wheat harvests only relate to the summer period and depend on both temperature and rainfall. This means that, while the extreme years stand out, many of the fluctuations cannot be attributed to a specific factor.

WINE HARVESTS

The wine-harvest date, when grapes were picked for wine-making, is a more reliable measurement of the temperature during the growing season in the northern hemisphere (April to September). The best-known record of wine-harvest dates has been built up from documents kept in parts of northern France where there is a centuries-old tradition of villages and towns declaring the local vineyards formally ready for harvesting. These records provide a good guide to summer temperatures between 1484 and 1879. Long-term fluctuations do, however, have to be viewed with caution. For instance, a period of late harvest dates in the late 18th century and early 19th century does not tally with available instrumental records for the period, and has more to do with changing fashions in the sweetness of wine: the later the harvest, the sweeter the wine.

FILLING THE GAPS

The principal limitation of many other historical records is that frequently they only note exceptional events. This means that many records are often exceedingly patchy, as they cover only a few years in any century. A good example is the record of polar bears coming ashore from ice floes around Iceland. On the face of it, accounts of polar bears in Iceland suggest that the Arctic pack-ice extended further south in the 13th and 14th centuries. But there are big gaps in the record, which mean we may be attaching too much importance to the fact that these rare events became a little more common after about 1300.

The way round this limitation is to find more records that may tell us about the climate at the time. An examination of all the extremes recorded can make it possible to build up a more complete picture of changes. An example of this is the work of Christian Pfister, at the University of Berne, Switzerland, on the changes in temperature and rainfall in Switzerland for each season since the early 16th century (see below). He combined written records of snowfall, freezing of rivers and floods with flowering dates of trees and plants, cereal and wine-harvest dates and yields, to provide a year-round measure of the climate.

Above: Harvesting was always a time of maximum effort when the weather was good – any deterioration in conditions could spoil the precious grain.

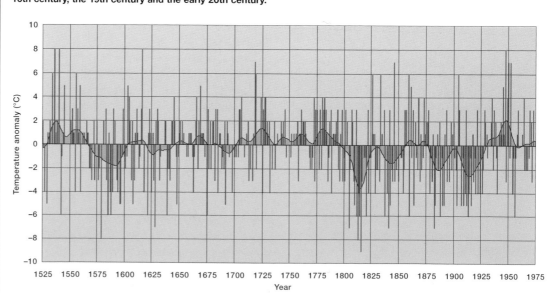

SWISS SUMMER TEMPERATURES
By using a variety of sources of information about the state of vegetation and harvest records, it is possible to estimate the temperature in Switzerland since the early 16th century. This record shows that the coldest periods were in the late 16th century, the 19th century and the early 20th century.

Left: An example of flooding in China in the early 20th century.

Below: Frantic efforts to stem a breach in the flood defences during the floods of August 1998 in Jiujiang, Hubei Province, China.

CHINA'S MASS OF HISTORICAL RECORDS

Another example is the pattern of droughts and floods in China that has been extracted from the huge store of historical records. These extend back over a thousand years and can provide detailed descriptions of variations in rainfall during the summer across this vast country. The references are extensive because, every year during the last five centuries, the Chinese authorities have graded the droughts and floods for over one hundred locations in the eastern part of the country.

In many years, the pattern of rainfall is fragmented, with plentiful rain in some parts of the country and drought in others. For example, in 1560, both the north and the south had drought while the Yangtze basin was flooded. The records include such items as: in Datong men were eating men, in Shijiazhung there was no rain in spring or summer and in Beijing there was a locust plague. In contrast, there were overwhelming floods in some other regions including Nanking and Yichang.

What emerges from this analysis is that during the summer half of the year the country is usually in the grip of one of about half-a-dozen basic weather regimes, such as droughts or floods. However, the records show no particular sequence for the occurrence of these regimes, or any marked change in the incidence of one or more of them over time. So, all in all, they suggest that the climate of this region has experienced the same considerable fluctuations for at least the last five centuries, and probably even longer.

As a rule, historical records can rarely be used to build up a real picture of climate change. Where the records are few and far between, great care has to be taken in not attaching too much importance to fluctuations in rare events. Only when there are frequent observations over a significant period can these records be used on their own to draw conclusions on climate change. Otherwise, their main value, in conjunction with proxy data, is that they provide background detail in assessing the social impact of climate variations and in building up a more complete picture of any change. Even if not strictly reliable in their own right, historical records do add colour to instrumental measurements and proxy data, and show how climate change has affected society and people.

Proxy records 40–1

Past wine harvests 103

PROXY RECORDS

THE KEY TO UNDERSTANDING PAST CLIMATIC CHANGE
IS FOUND IN A WEALTH OF SOURCES – FROM THE INSIDE
OF TREE TRUNKS TO THE BOTTOM OF OCEANS.

It is not always straightforward to build up a picture of what our climate used to be like in the past and how much it has changed. Instrumental and historical records are only available for a few hundred years of the Earth's 4.6 billion-year history. As such, climatologists have to turn to analysis of other weather-sensitive factors, which are known as proxy records – these include tree rings, ice cores, ocean sediments and pollen.

TREE RINGS

Tree rings form in the inside section of tree trunks and roots, with each ring indicating one year's growth. Climatologists studying tree rings are faced with the challenge of establishing precisely what combination of temperature, rainfall and soil-moisture levels brought about changes in ring thickness or changes in the wood structure within rings. They do this by observing tree-ring widths in a certain area for the period since standard meteorological measurements began. By comparing variations in the tree rings with the modern statistics of different weather conditions, such as summer temperature or annual rainfall, they are able to calculate how much of the ring variation can be attributed to specific climatic factors.

More detailed information is obtained from analysis of the structure of the wood within individual rings. Some factors, such as the density of the wood and the width of the rings, can reveal what the temperatures were like early and late in the season. For instance, stunted early growth could indicate late cold springs while lack of late growth could reflect cool short summers, and frost damage in late-spring growth is evidence of an exceptionally cold spell.

OCEAN SEDIMENTS

The sediments that collect on the beds of deep oceans provide a valuable source of data on long-term climate change. This ooze is formed largely of the bodies of the fossil shells of tiny creatures (foraminifera), which were living in either the surface or deep waters of the oceans. So extracting a core down through this sediment, which has collected over hundreds of thousands of years, provides a record of the species living in a particular area in the past. These sediments occur across all the major ocean basins and so it is possible to build up a picture of global changes from the results of a substantial ocean drilling programme, which has taken place over the last three decades.

Above: Variations in the thickness of annual tree rings can be used to draw conclusions about variations in the weather from year to year.

Right: A false-colour scanning electron micrograph of assorted pollen grains showing how the size, shape and surface characteristics differ from one species to another. By counting the number of species and their abundance in lake sediments and peat deposits, it is possible to analyse how the climate has varied over time.

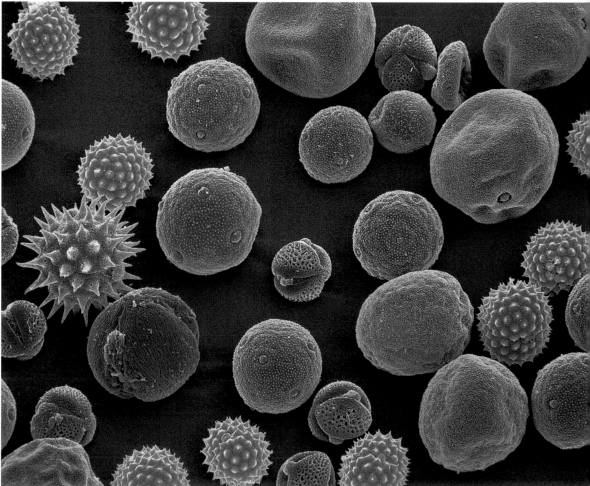

38–9 Historical records

The types of tiny animal species living in the surface waters at any given time, and their relative abundance, provide a guide to surface temperatures because some species only survive in warm or cold water. So, by mapping the populations of different creatures at different times using cores taken from around the world, it is possible to construct a picture of how the temperature of the oceans' surface waters have varied over time.

Ocean sediments can also give us an idea about the fluctuating extent of the ice sheets in the past. The proportion of light and heavy oxygen isotopes in the skeletons and shells of the foraminifera living in the deep water reveals how much water was, at one stage, locked up in the ice sheets. As the sheets grew, large amounts of the lighter isotope, rather than the heavier one, became locked in the ice. So as the extent of the ice sheets rose and fell, the ratio of the isotopes in the oceans changed accordingly. These results also reveal fluctuations in global temperature – the more area the ice sheets covered, the colder it was at lower latitudes.

When it came to dating cores extracted from the ice sheets, meteorologists originally assumed that the rate of sedimentation was constant – the same amount was laid down each year. However, when they compared how the isotope ratio rose and fell with the changes and wobble of the Earth's orbit, they decided that corrections were needed. This led to an accepted timescale, which reflects the major fluctuations in ice sheets over the last 500,000 years.

POLLEN RECORDS

Pollen from different species of trees and shrubs is often found in sediments at the bottom of shallow lakes and bogs. Because different plant species can only survive in distinct climatic conditions, their presence and abundance can be interpreted in terms of shifts in the local climate. Most pollens (from flowering trees and plants) and spores (mainly from ferns and mosses) are tiny: few exceed 100µm in diameter and the majority are around 30µm. Different species are identified by examining the size and shape of the outer wall of pollen cells (see left) and the number and distribution of apertures.

Most studies using pollen records have focused on changes in regional vegetation since the end of the last Ice age. Recent results have confirmed, however, that pollen records can be accurately dated back to the last interglacial, some 125,000 years ago, and can, in fact, provide detailed information about how the global climate has changed since the last Ice age. Furthermore, the close correlation between these results and those from the cores in European lakes, ice cores and ocean sediments means pollen records will play an increasing part in identifying global patterns of climate change during the last Ice age.

HOW CAN WE INTERPRET PROXY DATA?

Proxy data are rarely a direct measure of a single meteorological factor. For instance, the width of tree rings is determined by the temperature and rainfall over the growing season, as well as groundwater levels, which reflect rainfall in earlier seasons. Only where the trees are growing near their climatic limit (the extremes of climate in which they

Above: An example of a bristlecone pine, which lives in the White Mountains for up to 5000 years. Its longevity makes it invaluable for lengthy tree-ring studies.

survive) can most of the growth be attributed to a single factor, such as summer temperature. In order to draw climatic conclusions from other records (for example, the analysis of the pollen content or the creatures deposited in ocean sediments) we must first consider the sensitivity of plants and creatures to the climate and how their distribution is a measure of the climate at the time. When a variety of different proxy records is combined, however, it is possible to build up a surprisingly comprehensive picture of major environmental changes long ago.

ICE AGES

FOR MUCH OF ITS HISTORY LARGE AREAS OF THE EARTH'S SURFACE HAVE BEEN BURIED UNDERNEATH MANY THICK LAYERS OF ICE.

**MILUTIN MILANKOVITCH
(1879–1958)**
A Yugoslav geologist, Milankovitch spent 30 years trying to explain the occurrence of ice ages. He calculated variations in radiation striking different parts of the Earth due to changes in our planet's orbit and orientation, and tried to show that the changes in the amount of energy falling at different latitudes at different times of the year were enough to trigger glaciation.
At the time, evidence did not support this, but improved data in the 1950s and 1960s breathed new life into his theory, which is now regarded as the most probable explanation for the ice ages.

Below: The high reflectivity of snow and ice in polar regions makes them an important factor in climate change.

The ice ages have struck intermittently and have been separated by periods of relative warmth, which are known as interglacials. The interglacial that we are currently experiencing began about 17,000 years ago. Ice ages are thought to be caused by variations in the Earth's orbit, which affect how much sunlight reaches the planet at any one time.

EARLY PIONEERS OF ICE-AGE THEORY
The theory that ice sheets at one stage covered a large part of northern Europe was first proposed in 1795 by James Hutton, the founder of scientific geology, and was reiterated by a Swiss civil engineer, Ignaz Venetz, in 1821. But these ideas received scant attention until the summer of 1836, when the Swiss naturalist Louis Agassiz became convinced that blocks of granite in the Jura had somehow been transported at least 100km (62 miles) from the Alps. In 1837 he first coined the term ice age (Eiszeit) and in 1840 published his ideas in a ground-breaking book. At first, the theory was ridiculed by the geological community, but his passionate advocacy of the ice-age concept and a growing body of evidence prevailed.

Following the work of Agassiz, other geologists started to put forward theories about the length of past ice ages. They believed that over the last 600,000 years there had been four glacial periods lasting around 50,000 years, separated by warm interglacials ranging from 50,000 to 275,000 years in length. They thought that the present interglacial started about 25,000 years ago and that it was destined to last at least as long as previous interglacials and possibly indefinitely. Evidence of the stately progression of ice sheets was seen in glaciated landscapes of the northern hemisphere: scoured U-shaped valleys; eroded mountains; various mounds of debris that were carried and moulded by moving ice (glacial tills, terminal moraines and drumlins); ridges of gravel and sand formed by melting water flowing

out of the edge of ice sheets (eskers) and rocks so different in shape and content from those around them that must have been transported by ice (erratics).

RECENT DISCOVERIES
In the 1950s, analysis of ocean sediments from the tropical Atlantic and Caribbean revealed that there have been seven glacial periods during the last 700,000 years. The Italian geologist Cesare Emiliani suggested that ice ages struck the Earth roughly every 100,000 years. Subsequent studies of ocean sediments, pollen records from parts of Europe that were covered by ice and Antarctic ice cores have confirmed that Emiliani's conclusions were correct and have revealed that these cold periods were interspersed with shorter warm interglacials. Additionally, within each glacial period the climate has ranged from extreme cold to near-interglacial warmth.

EVENTS DURING THE LAST ICE AGE
The last Ice age started about 117,000 years ago and lasted for about 100,000 years. During the first 40,000 years, ice sheets slowly built up on the northern continents. Thereafter a series of changes of the most striking suddenness took place, as revealed in ocean sediment and ice-core data. In particular, the Heinrich layers in ocean sediment show that the icy Earth underwent a series of dramatic warmings. Named after the scientist who first identified them, these layers are thought to be made up of debris that was carried out into the North Atlantic by a surge of icebergs when part of the ice sheet covering North America collapsed suddenly. So the last Ice age was not a period of unremitting cold but of unsettled climate, which varied over the years and over the millennia.

The ice sheets reached their greatest extent about 18,000 years ago. At this time, ice up to 3km (2 miles) thick covered most of North America as far south as the Great Lakes, all of Scandinavia, the northern half of the British Isles and the Urals. In the southern hemisphere much of Argentina, Chile and New Zealand was under ice, as were the Snowy Mountains of Australia and the Drakensbergs in South Africa. The total amount of ice locked up in these ice sheets is estimated to be some 90 million km3 (22 million cubic miles) as compared to the current figure of about 30 million km3 (7 million cubic miles). This was enough to reduce the average global sea level by 90–120m (300–400ft).

The average global temperature at the height of the last Ice age was at least 5°C (41°F) and lower than current values. Over the ice sheets of the northern hemisphere it was 12–14°C (22–25°F) colder, and recent studies suggest the tropics may also have been as much as 5–6°C (9–11°F) cooler. Overall the global temperature was 8°C (14°F) lower. So there is still some scientific debate as to just how much colder it was during the coldest periods of the last Ice age. Then suddenly around 15,000 years ago a dramatic warming started, ushering in the more benign climate, known as the Holocene, which has been the feature of the last 10,000 years.

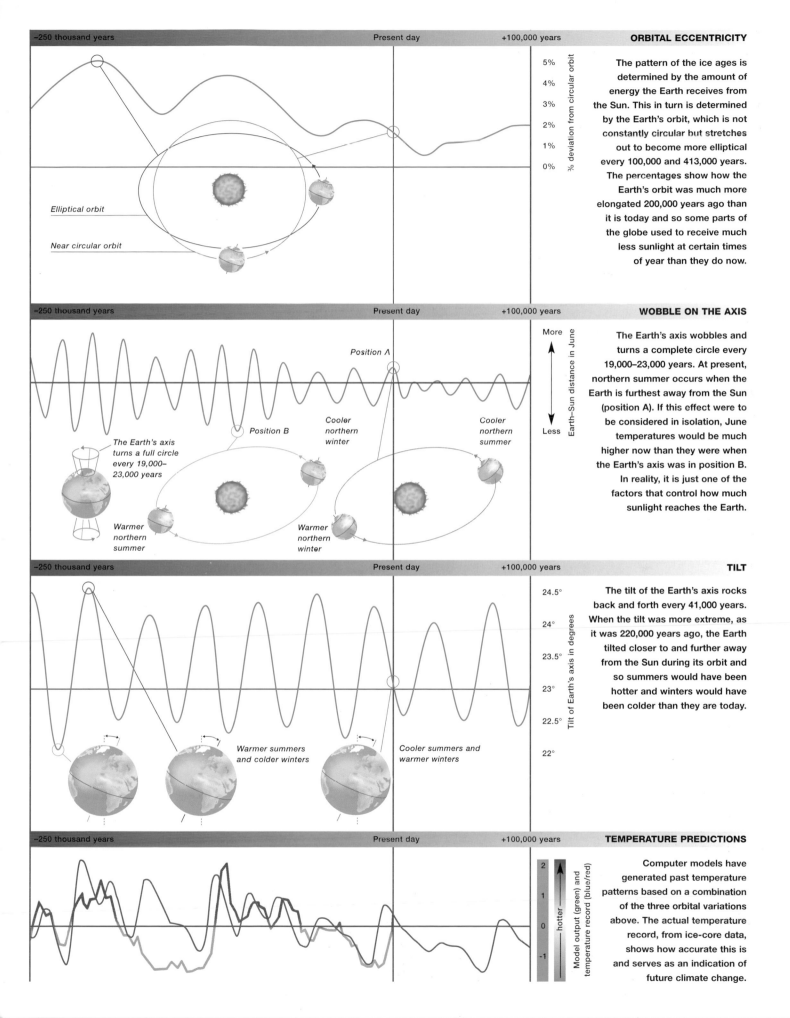

ORBITAL ECCENTRICITY

5%
4%
3%
2%
1%
0%

% deviation from circular orbit

The pattern of the ice ages is determined by the amount of energy the Earth receives from the Sun. This in turn is determined by the Earth's orbit, which is not constantly circular but stretches out to become more elliptical every 100,000 and 413,000 years. The percentages show how the Earth's orbit was much more elongated 200,000 years ago than it is today and so some parts of the globe used to receive much less sunlight at certain times of year than they do now.

Elliptical orbit

Near circular orbit

WOBBLE ON THE AXIS

More

Less

Earth–Sun distance in June

The Earth's axis wobbles and turns a complete circle every 19,000–23,000 years. At present, northern summer occurs when the Earth is furthest away from the Sun (position A). If this effect were to be considered in isolation, June temperatures would be much higher now than they were when the Earth's axis was in position B. In reality, it is just one of the factors that control how much sunlight reaches the Earth.

Position A

Position B

Cooler northern winter

Cooler northern summer

The Earth's axis turns a full circle every 19,000–23,000 years

Warmer northern summer

Warmer northern winter

TILT

24.5°
24°
23.5°
23°
22.5°
22°

Tilt of Earth's axis in degrees

The tilt of the Earth's axis rocks back and forth every 41,000 years. When the tilt was more extreme, as it was 220,000 years ago, the Earth tilted closer to and further away from the Sun during its orbit and so summers would have been hotter and winters would have been colder than they are today.

Warmer summers and colder winters

Cooler summers and warmer winters

TEMPERATURE PREDICTIONS

2
1
0
−1

hotter

Model output (green) and temperature record (blue/red)

Computer models have generated past temperature patterns based on a combination of the three orbital variations above. The actual temperature record, from ice-core data, shows how accurate this is and serves as an indication of future climate change.

CONTINENTAL DRIFT

THE EARTH'S SURFACE HAS NOT ALWAYS LOOKED THE WAY IT DOES TODAY. IT IS IN FACT CONSTANTLY MOVING AND CHANGING SHAPE.

Alfred Wegener, a German meteorologist, was the first to suggest in 1912 that the continents had at one stage been joined together in a supercontinent that he called Pangaea and somehow shifted to reach their current positions. But it was not until the 1960s that his visionary but unsatisfactory theory was developed into an accepted form.

HOW DID THIS AFFECT THE CLIMATE?

Continental drift has brought about slow changes in the climate. First, when two land masses collided, mountains were created – these generate a new climate and block or divert atmospheric circulation. Second, the shifting land masses altered global ocean circulation. Third, moving land masses meant changing amounts of vegetation, which brought about variations in the carbon dioxide levels in the atmosphere. Although geological changes seem immeasurably slow compared to the burning current issues of climate change, they provide important messages for understanding how that change initially came about.

THE PRECAMBRIAN AGE

We know very little about the first 90 per cent of the Earth's lifetime. We are not sure of where the oceans and the continents were, or the precise constituents of the atmosphere. The first forms of life, algae, came into existence about 3800 million years ago (mya) when it is believed that the climate was about 10°C (18°F) warmer than now. All we know is that between 2700 and 1800 mya, there were glacial conditions for at least part of the time.

The mists began to clear about 1000 mya. There was a glacial period lasting 200 million years in the late Precambrian Age that was marked by three ice ages, which reached their coldest peaks around 800, 700 and 590 mya respectively.

THE PALAEOZOIC ERA

About 700 mya there was a rapid growth in the number and diversity of developed forms of plants and animals with soft bodies. Then about 530 mya, during the Cambrian Age, there was another sudden dramatic acceleration in evolution when all the basic designs of animal life involving shells and hard skeletons made their first appearance in the fossil record. This upsurge in the evolutionary rate is often termed the Cambrian explosion, and it was at this stage that the Phanerzoic Era (our 'Age of Visible Life' covering the Palaeozoic, Mesozoic and Cenozoic Eras) began.

The climate became warmer during the Cambrian age and remained relatively mild for most of the period. During the next 250 million years or so the migration of the continents centred on Gondwanaland, which was made up of Africa, South America, India, Antarctica and Australia. Other continental fragments included parts of what are now North America and Eurasia. We do not know much about the climate of this period, but there was a brief glacial period at the end of the Ordovician and the start of the Silurian Eras. Evidence for this can now be found in rocks in the Sahara, which have ice marks on them suggesting that the region was close to the South Pole at this time.

During the Carboniferous Age the temperature dropped, culminating in the long Permo-Carboniferous glacial period from 330 to 250 mya. This icy epoch may have been the coldest period in the Earth's history. Its later stages coincided with the formation of the supercontinent, Pangaea, when all the Earth's land masses came together and stretched from the equator to the South Pole. Vast areas of this supercontinent were probably covered in ice, although it may have been warm elsewhere in the world.

THE MESOZOIC: AGE OF THE DINOSAURS

During the Mesozoic Era, stretching from 250 to 65 mya, Pangaea drifted towards lower latitudes and broke up. This was a period of generally warm climate, which is usually associated with the 'Age of the Dinosaurs'. A cooler period

---

PLATE TECTONICS

The theory of plate tectonics had a slow birth from around 1600 when the first explorers started to map the coastlines of Africa and South America and noted that, when these two coastlines were seen on paper, they fitted together like two pieces of a jigsaw. But in every other sense, the idea seemed preposterous and attracted ridicule from thinkers and scientists for generations. It was not until the 1950s that it became clear that the surface of the Earth is not a single unit but is split into a number of plates, which move at a rate of several centimetres per year in relation to each other. Tectonic activity is also responsible for earthquakes and volcanoes.

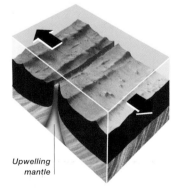

Upwelling mantle

SPREADING
In some places adjacent plates are pulled apart and new oceanic crust is pulled up to fill the gap that has been created.

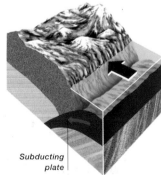

Subducting plate

SUBDUCTION
If land and ocean collide, oceanic crust is dragged beneath the continent and absorbed back into the Earth's interior.

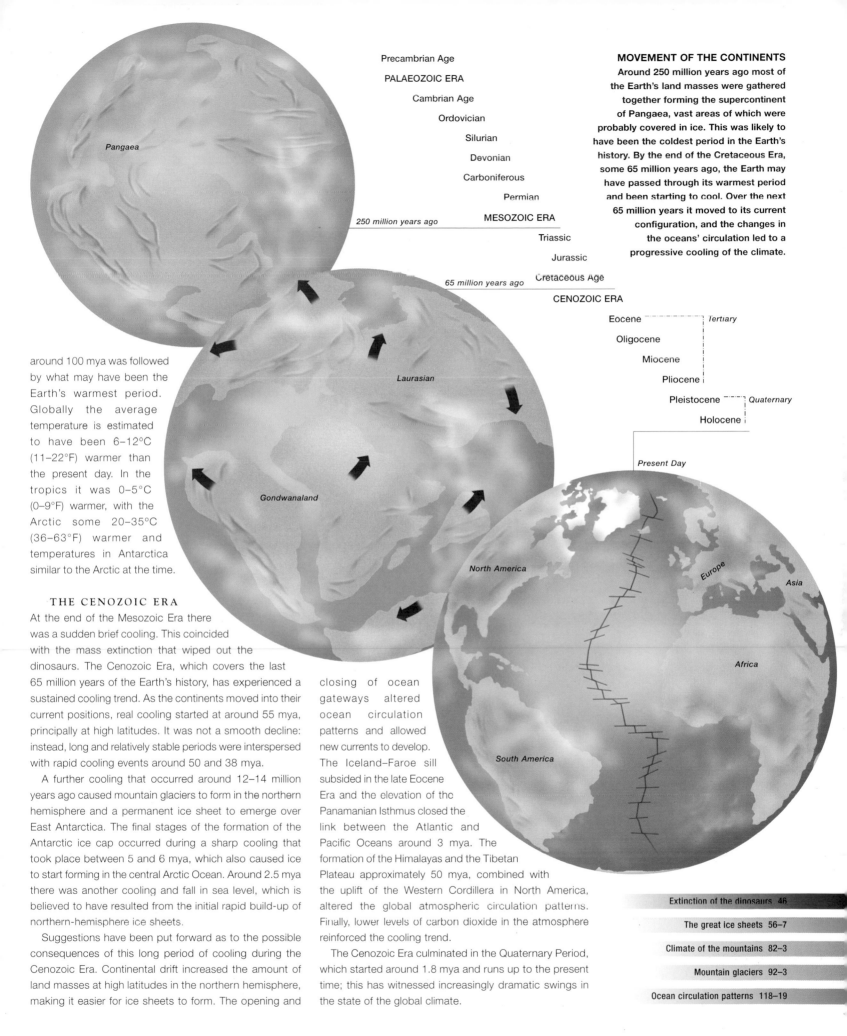

Precambrian Age

PALAEOZOIC ERA

Cambrian Age

Ordovician

Silurian

Devonian

Carboniferous

Permian

250 million years ago **MESOZOIC ERA**

Triassic

Jurassic

65 million years ago Cretaceous Age

CENOZOIC ERA

Eocene ------- Tertiary

Oligocene

Miocene

Pliocene

Pleistocene ----- Quaternary

Holocene

Present Day

MOVEMENT OF THE CONTINENTS
Around 250 million years ago most of the Earth's land masses were gathered together forming the supercontinent of Pangaea, vast areas of which were probably covered in ice. This was likely to have been the coldest period in the Earth's history. By the end of the Cretaceous Era, some 65 million years ago, the Earth may have passed through its warmest period and been starting to cool. Over the next 65 million years it moved to its current configuration, and the changes in the oceans' circulation led to a progressive cooling of the climate.

around 100 mya was followed by what may have been the Earth's warmest period. Globally the average temperature is estimated to have been 6–12°C (11–22°F) warmer than the present day. In the tropics it was 0–5°C (0–9°F) warmer, with the Arctic some 20–35°C (36–63°F) warmer and temperatures in Antarctica similar to the Arctic at the time.

THE CENOZOIC ERA

At the end of the Mesozoic Era there was a sudden brief cooling. This coincided with the mass extinction that wiped out the dinosaurs. The Cenozoic Era, which covers the last 65 million years of the Earth's history, has experienced a sustained cooling trend. As the continents moved into their current positions, real cooling started at around 55 mya, principally at high latitudes. It was not a smooth decline: instead, long and relatively stable periods were interspersed with rapid cooling events around 50 and 38 mya.

A further cooling that occurred around 12–14 million years ago caused mountain glaciers to form in the northern hemisphere and a permanent ice sheet to emerge over East Antarctica. The final stages of the formation of the Antarctic ice cap occurred during a sharp cooling that took place between 5 and 6 mya, which also caused ice to start forming in the central Arctic Ocean. Around 2.5 mya there was another cooling and fall in sea level, which is believed to have resulted from the initial rapid build-up of northern-hemisphere ice sheets.

Suggestions have been put forward as to the possible consequences of this long period of cooling during the Cenozoic Era. Continental drift increased the amount of land masses at high latitudes in the northern hemisphere, making it easier for ice sheets to form. The opening and

closing of ocean gateways altered ocean circulation patterns and allowed new currents to develop. The Iceland–Faroe sill subsided in the late Eocene Era and the elevation of the Panamanian Isthmus closed the link between the Atlantic and Pacific Oceans around 3 mya. The formation of the Himalayas and the Tibetan Plateau approximately 50 mya, combined with the uplift of the Western Cordillera in North America, altered the global atmospheric circulation patterns. Finally, lower levels of carbon dioxide in the atmosphere reinforced the cooling trend.

The Cenozoic Era culminated in the Quaternary Period, which started around 1.8 mya and runs up to the present time; this has witnessed increasingly dramatic swings in the state of the global climate.

ASTEROID IMPACT

FEW CLIMATIC EVENTS STRETCH THE IMAGINATION
MORE THAN THE POSSIBILITY OF THE EARTH BEING
HIT BY A LARGE EXTRATERRESTRIAL OBJECT.

Although many scientists regard asteroid impacts as being fixed in the realms of science fiction, an object as small as 1km (0.6 mile) wide would have a massive temporary impact on the climate.

RECENT COLLISION COURSES
The most recent collision occurred on 30 June 1908 in Siberia when either a meteorite or a fragment of a comet measuring some 50m (160ft) wide exploded in the atmosphere about 10km (6 miles) over the Tunguska River. Its force was equivalent to a bomb of 10–20 megatons of TNT and it devastated 2000km² (800 square miles) of forest (see right). While the climatic impact of this event was small, a larger object might have more serious consequences.

In July 1994, fragments of the comet Shoemaker–Levy 9 collided with the giant planet Jupiter, providing a graphic illustration of the potential consequences if such a large

object were to strike the Earth. The biggest fragment, about 3–5km (2–3 miles) wide, ejected gas to a height of 1300km (800 miles) and created a cloud larger than the entire Earth.

EXTINCTION OF THE DINOSAURS
Collisions of this sort may have caused some of the major extinctions of species in the past. A huge asteroid impact is believed to have brought about the extinction of the dinosaurs at the end of the Cretaceous Age, 65 million years ago. The Yucatán Peninsula in Mexico is thought to have been the site of this impact, where an object some 10km (6 miles) wide formed a crater measuring 200–300km (125–190 miles) in diameter. The immediate consequences of this impact must have been cataclysmic – a combination of massive earthquakes, huge tidal waves, widespread forest fires and vast amounts of dust and debris hurled high into the atmosphere that would have done untold damage.

ASSESSING POTENTIAL IMPACTS
Estimates of the environmental consequences of an asteroid impact are usually calculated in terms of the equivalent atomic bomb, as both would produce clouds

Above: Some 2000km² (800 square miles) of forest was damaged by the Tunguska event in west Siberia in June 1908.

of dust and debris that would have similar effects on our atmosphere. If a 1km (0.6 mile) meteor were to collide with the Earth, it would give off energy equivalent to 100,000 megatons, while a 10km (6 mile) object, comparable to the one that hit 65 million years ago, would be a thousand times bigger. In contrast, computer models show that, if the world's entire nuclear arsenal were to explode, it would give off a mere 6000 megatons, which gives us some idea of the scale of the impact of the prehistoric catastrophes.

Studies based on nuclear explosions have been carried out to estimate what the damage would be were the Earth to collide with a large extraterrestrial object. The land surface temperature would drop by between 20 and 40°C (36 and 72°F) within a matter of days. However, unless impacts were to occur on a regular basis, these changes would only be short-lasting.

CAN WE PREDICT THE NEXT IMPACT?

Estimates of the rate of impacts are based on the analysis carried out by Eugene Shoemaker on the Moon's craters (see right). This suggests that an event similar to the one that took place in Tunguska might happen every few hundred years, while a climatically significant impact of about 1000 megatons may occur every 10,000 years or so.

We do not know whether collisions are random or exhibit some regularity because it depends on the amount of debris near our solar system, and in Earth-crossing orbits in particular. This is why the National Space Administration (NASA) is financing a project for ground-based telescopes to search for asteroids and comets with a diameter of more than 1km (0.6 mile) that may be on a collision course with the Earth. Although we do not yet know how to divert any such objects, one thing is certain: if we don't know they are on their way, there is nothing we can do about them.

Left: Meteor Crater in north-east Arizona is about 50,000 years old. At 800m (2600ft) wide and 200m (650ft) deep, it is the clearest example of an asteroid impact on the Earth.

EUGENE SHOEMAKER (1928–97)
Eugene Shoemaker founded the scientific study of impact craters on the Earth, Moon and other planets and their satellites, as well as pioneering surveys of near-Earth asteroids and comets, often in collaboration with his wife Carolyn. While studying volcanic rocks in south-west USA in the late 1940s he became interested in whether the craters on the Moon were volcanic or caused by impacts. After visiting Meteor Crater in Arizona (see left), he became convinced that both it and lunar craters were formed by impacts. His subsequent work mapping craters caused by nuclear explosions at the Nevada test site provided further evidence of the nature of the impact process. His association with the Apollo programme enabled him to develop a method of dating the craters by measuring their density. The study of Moon rocks also provided confirmation of the impact origin of lunar craters. In 1972 he began a programme to search for near-Earth asteroids. Over 25 years he, his wife and Eleanor Helin discovered 140 such asteroids, and some 32 comets were discovered that bear the Shoemaker name. His most famous achievement was, with his wife and amateur comet discoverer David Levy, the discovery in 1993 of the Comet Shoemaker–Levy 9, which hit Jupiter in 1994 with such spectacular results.

POLAR REGIONS

CLIMATE OF THE ARCTIC

THICK LAYERS OF ICE, COVERING LAND AND SEA, ARE THE PERVADING FEATURES OF THE POLAR LANDSCAPE AND CONTROL THE CLIMATE OF THESE REGIONS.

The region north of the Arctic circle encompasses a variety of climatic conditions. However, all these northern wildernesses feature extremes of cold and storminess, and surprisingly low levels of precipitation.

AN ICY DESERT

Data about climatic conditions in the Arctic circle come from polar expeditions, including the epic voyage of Nansen and the *Fram* (see far right). As a result, they do not represent the type of continuous climatic record that is normally used to provide a basis for examination and comparison. Nevertheless, they do provide a clear picture of the severe conditions that are experienced in this northerly part of the world.

From November to March the average temperatures in the Arctic region fall to between –25°C (–13°F) and –35°C (–31°F), with lowest figures of around –50°C (–58°F). In the height of summer, the temperature, which is controlled by the all-pervading influence of the slowly melting ice, hovers just below freezing. The highest temperatures, which occur in the continual sunshine of mid-summer, are only ever a few degrees above freezing. Although the Arctic is wracked with frequent storms and strong winds, the total annual precipitation is very low, which is why the region is often referred to as an 'icy desert'.

Above: The Arctic pack-ice is always moving and so it gets pushed and pulled every which way. This forces pressure ridges to rise upwards and the ice cover to break in places, as this sunlit view shows.

POLAR SEA ICE

Most of the Arctic Ocean is covered by a layer of frozen sea ice, several metres thick, which is constantly changing shape as the stresses of the wind above it, and the much more gradual ocean circulation below it, push and pull in all directions. As such, local conditions vary greatly with shifts in the ice cover, depending on whether the ice is piled up in large hummocks (low mounds of ice) and pressure ridges (crumples where two ice blocks meet) or stretched out to form leads (breaks of open water).

Minimum extent of sea ice

Maximum extent of sea ice

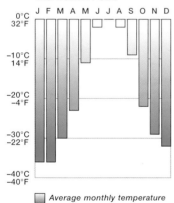

THE NORTH POLE

Average monthly temperature

Negligible precipitation

Right: There are many breaks in the Arctic sea-ice cover, especially at the outer limits in the summer. So getting around means taking to the water, as this polar bear is ably demonstrating.

16–17 Oceans and currents

THE JOURNEY OF THE FRAM

The Norwegian artist, statesman, humanitarian, zoologist, oceanographer and intrepid Arctic explorer, Fridtjof Nansen, was a remarkable man. He achieved fame in 1888–9 when he crossed Greenland from east to west and led the Norwegian delegation to the League of Nations. After World War I, he directed a programme to bring relief to famine-stricken Russia in 1921 and was awarded the Nobel Peace Prize in 1922.

Nansen's great contribution to meteorology was to convince a sceptical world that a ship could be transported across the Arctic Ocean in the ice if it entered at the right place. In 1893 he sailed from Oslo in a ship named *Fram* and entered the ice 350km (218 miles) north of Ostrov Kotel'nyy with the objective of being carried in the ice over the North Pole to emerge somewhere between Greenland and Spitzbergen after two or three years. By 1895 Nansen realized the ship would not pass over the Pole, and decided to leave it and set off by sledge for the Pole with his friend, Hjalmar Johansen. After an epic voyage, during which they reached 86°N, they returned to Franz Josef Land, and spent the winter there in 1895–6. Then, in May 1896, stranded and with no means of getting home, by an astonishing coincidence they met the English explorer, Frederick Jackson, and returned with him to Norway in a ship called *Windward*. The *Fram* and its crew emerged safely from the ice north-west of Spitzbergen and reached Tromsø a week after Nansen's return. This whole audacious venture provided a huge amount of information about conditions in Arctic regions and the large-scale movements of the ocean.

Above: Fridtjof Nansen (1861–1930) whose ship, the Fram, was transported across the Arctic Ocean.

The ice follows a roughly elliptical path westwards through the Beaufort and Chukchi Seas and then swings northwards over the North Pole and back towards Ellesmere Island. It moves at a speed of a kilometre or two a day. Some of this ice escapes from the Arctic down the east of Greenland, and some exits via the Bering Straits.

THE GREENLAND ICE SHEET

The high levels of the Greenland ice sheet are the coldest places in the northern hemisphere, exceeded only by the equivalent regions of Antarctica. At altitudes of around 3km (10,000ft) the average annual temperature is about –30°C (–22°F), ranging from –40°C (–40°F) or below between November and March to nearly –12°C (10°F) during mid-summer. In spite of the intense cold, the snowfall is not high and only forms about 50cm (20in) of ice each year. But over the centuries this has built up a massive ice sheet, which provides a permanent record of the weather conditions over tens of thousands of years.

WARM WATER AND POLAR LOWS

The unremitting cold of high latitudes is lessened when warm water is carried up into the Arctic basin from the North Atlantic. This pushes the sea ice back and means that Spitzbergen, Norway, at 78°N in winter, is not as cold as Winnepeg, Canada, at 49°N, or Kasalinsk in Central Asia at 45°N. However, in summer the temperature only creeps up to around 5°C (40°F). The average precipitation is about 35cm (14in) in Spitzbergen, which is considerably higher than most other polar regions.

Another feature of the regions close to the edge of the sea ice is the spawning of small but intense low-pressure systems. Known as polar lows, these systems, even in their smallest form, have remarkably organized circulation with

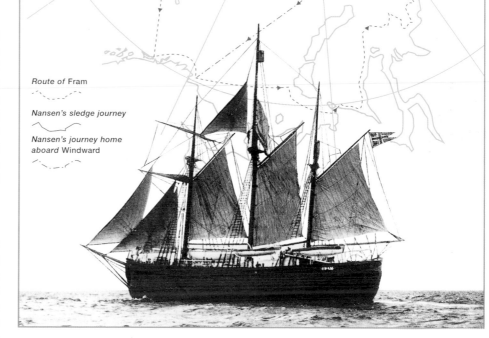

Route of Fram

Nansen's sledge journey

Nansen's journey home aboard Windward

a central 'eye'. Although the pressure drop across the systems is only very small, they can cause winds of up to 110kph (70mph) with driving snow and zero visibility. They can also become embedded in strong northerly outbursts, which occur in winter between Greenland and Norway, bringing heavy snow to northern Europe and the Baltic.

CLIMATE OF ANTARCTICA

ANTARCTICA IS THE COLDEST, MOST REMOTE
CONTINENT ON EARTH – ITS HUGE ICE SHEETS
CONTAIN 90 PER CENT OF THE PLANET'S FRESHWATER.

Sitting directly over the South Pole, the approximately circular symmetry of Antarctica exerts powerful influences on its own climate and that of much of the southern hemisphere.

GEOLOGICAL HISTORY

The Antarctic continent first became isolated about 40 million years ago. The Drake Passage opened up between South America and Antarctica, and the movement of Australia further north allowed the circumpolar current to develop, which isolated Antarctica in terms of climate. Ice sheets started to build up rapidly on the mountains of Antarctica around 29 million years ago but there is evidence that there were still areas of forest on the continent as late as 25 million years ago. However, for the last 5 million years or so the Antarctic continent has been covered by a huge ice sheet; this has dominated the climate of much of the southern hemisphere and controlled the evolution of flora and fauna on the continent and in the surrounding waters.

THE COLDEST PLACES IN THE WORLD

The near-circular symmetry of the ice sheets, which rise to altitudes of more than 3km (10,000ft), dominate the climatic conditions of the continent. High on the most remote parts of the East Antarctic ice sheet (often termed the 'pole of inaccessibility') are the coldest places on the Earth's surface. At the Vostok research station, which is run by the Russians at an altitude of 3420m (11,218ft), the average annual temperature is –55°C (–67°F), and never goes above about –20°C (–4°F), whilst the lowest figure ever recorded was a numbing –89.2°C (–128.6°F) on 21 July 1983. Even at the somewhat lower levels of the Amundsen–Scott Base at the South Pole, at an altitude of 2880m (9447ft), the average temperature is –49°C (–56°F). In spite of the storms and frequent fierce winds, snowfall amounts are only about 50cm (20in) per year – equivalent to a mere 50mm (2in) of rain.

Around the fringes of the continent conditions are marginally more hospitable. Where the sea ice largely disappears (eg Mawson at 67°S, 62°E) the average temperature in mid-summer rises to around freezing point, with occasional balmy days peaking at 4 or 5°C (40°F), but in winter the average temperatures are about –18°C (0°F) and the frequent storms plus almost continual darkness make it an extremely harsh environment. In addition, icy

winds gusting off the ice sheet (katabatic winds) are a common feature of these coasts. They form a shallow layer of intensely cold surface air that flows downslope, sometimes at great speeds and, together with local winds, they whip lying snow into a blizzard. These winds regularly exceed 90kph (55mph) and can gust at over 200kph (125mph). A measure of the extraordinary intensity of the winds is gained from the observations by Douglas Mawson's 1912–13 expedition at Cape Dennison. During the 12-month period, the average wind speed exceeded 65kph (40mph) on two-thirds of the days. Annual precipitation figures in these peripheral regions is equivalent to about 500mm (20in) of rainfall.

THE SOUTHERN OCEANS

The symmetrical shape of Antarctica (combined with the absence of other land masses at high latitudes) imposes its stamp on the climate of the southern oceans. The strong circumpolar current – a consequence of the symmetry of Antarctica – reduces the transport of warm surface water from lower latitudes, which means the strong westerly winds and continual string of low-pressure systems are

Permanent ice shelf

Minimum summer extent of sea ice

Maximum extent of sea ice

THE SOUTH POLE

J F M A M J J A S O N D

0°C / 32°F
–10°C / 14°F
–20°C / –4°F
–30°C / –22°F
–40°C / –40°F
–50°C / –58°F
–60°C / –76°F

■ *Average monthly temperature*

Negligible precipitation

16–17 Oceans and currents

44–5 Continental drift

50–1 Climate of the Arctic

Left: Born at the end of the long Antarctic winter, these Emperor penguin chicks wait, under watery skies at Atka Bay on the Weddell Sea, for adults to return with food.

Below: A composite infrared satellite image of Antarctica with snow shown as white, ice as white/blue and the exposed rocks of the Transantarctic Mountains as black.

fuelled mainly by the outward flow of cold air from the icy continent. The positions of these systems move relatively little during the year but do reflect the contracting and expanding pulse of the sea ice.

These conditions mean that the westerly winds are much stronger and more constant than they are elsewhere in the world. In winter the average wind speed, between 35 and 60°S, is more than twice the peak speed in the northern hemisphere at around 50°N. Furthermore, while the region of strong winds narrows in summer, the fastest speeds, if anything, increase at around 50°S. So the evocative epithets 'Roaring Forties' and 'Screaming Fifties', which were coined by the mariners on the 19th-century clippers and endorsed by present-day round-the-world sailors, are fully justified.

INVASION OF A FROZEN WILDERNESS

Life in Antarctica has evolved in spite of the harsh conditions found in the icy wastes of its ice sheet. Fish can thrive in waters that are –1.9°C (28.6°F) because they have a powerful form of antifreeze in their blood, and Emperor penguins can incubate their eggs through the winter even

though temperatures can fall to –55°C (–67°F). Unique microbes, which have evolved in isolation over the last 40 million years, also give an insight into the process of evolution. The survival of this pristine and beautiful wilderness may now be compromised, however, by human activities and global warming.

The scientific community, which occupies 50 bases around the continent, is committed under the Antarctic Treaty of 1959 to protect the environment. The last teams of huskies, for example, whose ancestors used to pull the sledges of the early Antarctic explorers, were flown out in 1994. However, there is growing evidence that human influence is beginning to threaten the local wildlife. Alien grasses now grow near some of the bases, stimulated in part by the increase in temperatures. Boats have brought their own problems to the region: rats are now endemic on some of the islands and diseases common in poultry appear to be killing penguins.

The threat posed by tourism is not yet as serious as that of scientific work, but there is a growing number of cruises that now venture into the icy region and the risks to this unique environment are likely to increase.

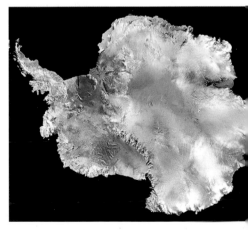

AURORAE

FEW NATURAL PHENOMENA ARE AS BREATHTAKING AS AURORAE – MOVING SHEETS OF COLOUR THAT PASS ACROSS OUR NIGHT SKIES.

Aurorae have fascinated sky watchers for many centuries. As long ago as the 1st century AD, the Roman emperor Tiberius sent his fire engines to the town of Ostia thinking that it was on fire because a spectacular aurora turned the entire sky red. Later, in medieval times, they were thought to be bad omens and often featured in folklore and mythology. Today, we know that aurorae are, in fact, a visible manifestation of the strange properties exhibited by our upper atmosphere.

HOW TO SEE AURORAE

Aurorae take place in both hemispheres – *aurora borealis* (northern lights) and *aurora australis* (southern lights). They can be seen to their best effect on cloudless nights away from artificial light. The best places to see aurorae are near the Arctic or Antarctic circles, but they can occasionally be seen at latitudes as low as 40°N or S. The frequency of observations, however, falls off sharply at lower latitudes, so they can be seen about 10 times more often at 60°N or S than at 50°N or S.

Aurorae can take a number of forms: firstly, they glow brightly in parts of the sky that are normally dark; secondly they form glowing arcs, rather like flattened rainbows; and thirdly they take the shape of a single or a number of rays extending upwards, sometimes from a glowing arc aurora.

Not to be confused with aurorae are noctilucent clouds – another beautiful meteorological phenomenon that occurs in the upper atmosphere. These are visible in summer at high latitudes and appear to their best advantage some time after sunset. Their brilliant colours are created by the sunlight reflected from ice crystals that form on meteoric dust at high levels.

LINKS WITH SUNSPOT CYCLES

The number of aurorae that occur over the Earth is dependent on the amount of activity that takes place on the Sun. The best opportunity of seeing an aurora is several days after a major solar storm. These storms tend to vary in frequency over the years but the best-known pattern is the 11-year cycle in sunspot numbers. The incidence of aurorae reaches a maximum a year or two after the number of sunspots peaks. The last sunspot maximum was between 1989 and 1991 and so we can expect the best opportunities to see the northern and southern lights around 2001 and then again about 2012.

Recent satellite studies have revealed that, while ground-level observations of the most intense aurorae do show the well-known link with solar activity, this relationship does not apply to all auroral phenomena. While the most intense aurorae observed in the Earth's dark zones can be linked with sunspots, this is cancelled out by other auroral activity over the sunlit part of the globe. This unexpected result shows that we must be cautious when linking specific changes in the upper atmosphere with climatic events that are taking place at the surface.

Above: The spectacular red and blue of the northern lights (aurora borealis).

Right: The southern lights (aurora australis) were photographed by the crew of the Shuttle Discovery in 1991 as part of a study of the characteristics of aurorae.

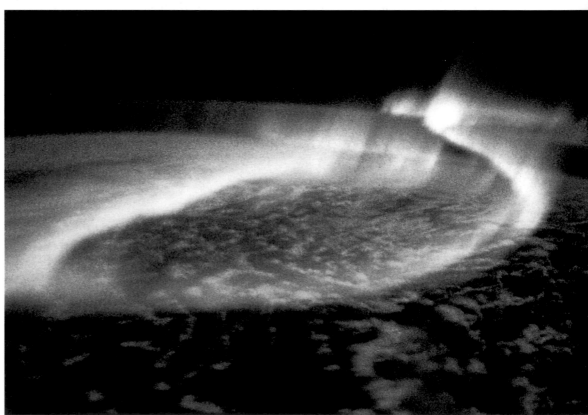

THE UPPER ATMOSPHERE AND AURORAE

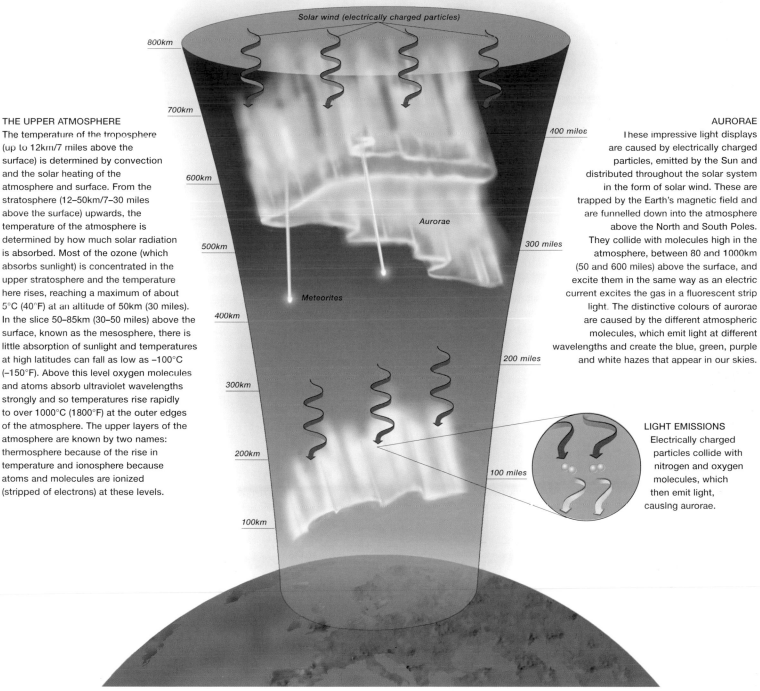

Solar wind (electrically charged particles)

800km

700km

400 miles

600km

Aurorae

500km

300 miles

Meteorites

400km

200 miles

300km

200km

100 miles

100km

THE UPPER ATMOSPHERE

The temperature of the troposphere (up to 12km/7 miles above the surface) is determined by convection and the solar heating of the atmosphere and surface. From the stratosphere (12–50km/7–30 miles above the surface) upwards, the temperature of the atmosphere is determined by how much solar radiation is absorbed. Most of the ozone (which absorbs sunlight) is concentrated in the upper stratosphere and the temperature here rises, reaching a maximum of about 5°C (40°F) at an altitude of 50km (30 miles). In the slice 50–85km (30–50 miles) above the surface, known as the mesosphere, there is little absorption of sunlight and temperatures at high latitudes can fall as low as –100°C (–150°F). Above this level oxygen molecules and atoms absorb ultraviolet wavelengths strongly and so temperatures rise rapidly to over 1000°C (1800°F) at the outer edges of the atmosphere. The upper layers of the atmosphere are known by two names: thermosphere because of the rise in temperature and ionosphere because atoms and molecules are ionized (stripped of electrons) at these levels.

AURORAE

These impressive light displays are caused by electrically charged particles, emitted by the Sun and distributed throughout the solar system in the form of solar wind. These are trapped by the Earth's magnetic field and are funnelled down into the atmosphere above the North and South Poles. They collide with molecules high in the atmosphere, between 80 and 1000km (50 and 600 miles) above the surface, and excite them in the same way as an electric current excites the gas in a fluorescent strip light. The distinctive colours of aurorae are caused by the different atmospheric molecules, which emit light at different wavelengths and create the blue, green, purple and white hazes that appear in our skies.

LIGHT EMISSIONS

Electrically charged particles collide with nitrogen and oxygen molecules, which then emit light, causing aurorae.

CAN AURORAE AFFECT THE CLIMATE?

The altitudes at which aurorae are formed are far too high to have any direct influence on the climate. However, some meteorologists argue that aurorae can influence the upper atmosphere, which in turn brings about changes in the lower atmosphere and climate. There is a fundamental problem with any theories that suggest that changes in the Earth's upper atmosphere, such as variations in the influx of electrically charged particles or cosmic rays from the Sun, can influence the climate. This rests in the fact that the Earth's lower atmosphere essentially acts as a heat engine driven by the solar radiation it absorbs and reradiates. This incoming and outgoing energy controls the state of the Earth's climate and will not be affected by tiny changes in high levels of the atmosphere.

This basic problem can only be overcome if the influx of electically charged particles or cosmic rays in some way alters the upper atmosphere enough to interfere with how much solar radiation actually reaches the Earth's surface. This means that in order to link the climate with fluctuations in solar activity, as seen in the spectacular displays of aurorae, meteorologists are faced with the challenge of proving that the variations actually alter the chemistry or cloudiness of the upper atmosphere to such an extent that there is a significant change in how much energy enters the atmospheric heat engine. So far, this challenge has not been met by scientists who wish to convince the climatological community that extraterrestrial changes in the flux of energetic particles into the upper atmosphere are an important part of global climate change.

Evidence of cycles 148

THE GREAT ICE SHEETS

ICE SHEETS ARE ENORMOUS DOMED GLACIERS THAT
COVER SUBSTANTIAL EXPANSES OF LAND, SUCH AS
ANTARCTICA AND GREENLAND.

Formed from centuries of snowfall, ice sheets hold the key to detailed information about past climatic change. Two other forms of glacier make up the familiar sights of polar regions: ice shelves, which project into the sea but remain attached to the land; and icebergs, which break away from ice sheets or ice shelves and float in the sea. All these glaciers are a major component of the hydrological cycle and exert a considerable influence on the climate.

GROWTH AND DECAY OF ICE SHEETS
The formation of ice sheets is a simple process. Water vapour from the oceans at low latitudes is transported by weather systems to higher latitudes. Some of it rains down along the way but is partially replenished by evaporation from the oceans. If the air then passes over Antarctica or Greenland, most of the remaining water vapour precipitates as snow, which then compresses over time to form ice. With glacial slowness, it then creeps back to the sea, where large blocks of ice are released as icebergs.

The issue of whether the ice sheets are currently growing or declining can only be addressed in terms of the most recent measurements available. Satellite observations since the mid-1980s are equivocal – the heights of the central ice sheets in both Antarctica and Greenland have showed no noticeable change. Recent aircraft measurements of the south-eastern fringes of the Greenland ice sheet do, however, suggest significant losses, which in places exceed 1m (3ft) per year and may well be an early sign of the impact of global warming.

Although only 2.1 per cent of the Earth's water is locked up in these ice sheets, fluctuations in their size alters global sea levels and leads to subtle changes in the properties of ocean water. These changes can be detected by studying the skeletons of tiny sea creatures that were preserved in the ocean sediments. At the same time, the climatic conditions can be deduced by examining the different layers of snow built up in the ice sheets (see below right).

ICEBERGS
It is estimated that 10,000–15,000 icebergs break off each year from glaciers along the coast of West Greenland. A few are carried as far south as 42°N by the Labrador Current. The size and shape of icebergs are the subject of formal classification ranging from the smallest, a 'growler', which is 5m (17ft) tall and 15m (50ft) long to the largest, the aptly named 'very large berg', which is 75m (240ft) tall and 230m (750ft) long. The iceberg season usually extends from March to June and since 1946 it has had an average length of 130 days. The longest seasons in recent decades were 1972 and 1973, both of which lasted 189 days. In spite of the recent warming, 1991 was also a long season, lasting 183 days from 23 February to 24 August.

USING ICEBERGS AS SCIENTIFIC BASES
In the Arctic Ocean huge icebergs, up to 25 x 30km (16 x 18 miles), break off the ice shelves of Ellesmere Island and Greenland and float around embedded in the frozen sea ice for a number of years. These ice islands are used as bases for scientific research because their great thickness and extent make them much safer than trying to work on the more variable sea ice. Moreover, the way in which they move provides useful information about the dynamics of the ocean. The first of these camps was

Above: Crevasses on a glacier in north-west Greenland provide evidence of the glacier's movement.

Right: A huge tabular iceberg that is situated off Adelaide Island in the Antarctic Peninsula.

A KEY TO PAST CLIMATIC CHANGE

Water molecules (H_2O) contain two forms of oxygen: the abundant lighter oxygen isotope ^{16}O and the rarer heavy isotope ^{18}O. In normal water, 99.76 per cent is the lighter version $H_2{}^{16}O$ and 0.2 per cent the heavier version $H_2{}^{18}O$, with other less common forms making up the rest. In cold temperatures, the heavier isotope ^{18}O is more likely to precipitate out of water vapour than the lighter isotope. Thus, by measuring the amount of ^{18}O in the ice sheets, meteorologists can estimate the temperature at which the water originally precipitated out. However, the quantity of ^{18}O is not a precise measurement because the water vapour reached the poles in a very convoluted way (see below). So, while change in the proportion of ^{18}O provides a good indication of how temperature in the region has changed over time, it is not an absolute measurement and even relative figures must be treated with care.

established by the Russians in 1938, and they have been a regular feature of North American and Russian research into the Arctic Ocean ever since.

ICEBERGS AND SHIPPING SAFETY

However useful icebergs may be for science, they pose a major hazard to shipping when released into the sea. The sinking of the Titanic in 1912 is a graphic example of this and had a profound influence on thinking about sea safety. It hastened the introduction of radio communications and transformed the attitude of sailors to the weather. It also galvanized international action to set up warning systems for ice conditions in the western North Atlantic. These hinge on ice patrols, maintained by the American Coast Guard since 1914, except during World War II.

The huge tabular icebergs that break off from the ice shelves around Antarctica, notably the Ronne–Filchner Ice Shelf in the Weddell Sea, do not constitute a hazard to shipping in the way icebergs do in the North Atlantic. Their size is, however, impressive and they can be as long as 90km (56 miles) and stand nearly 50m (165ft) out of the water. They may survive for several years, slowly breaking up as they move to lower latitudes.

In an extreme instance in 1894 an iceberg was recorded as far north as 26°S. The formation of huge icebergs in recent years, principally caused by the break-up of the Larsen B Ice Shelf in Antarctica, has been linked with global warming.

In the longer term there is an important question as to the role the ice shelves in the Ross and Weddell Seas play in the stability of the West Antarctica Ice Sheet and in the much talked about issue of rising sea levels.

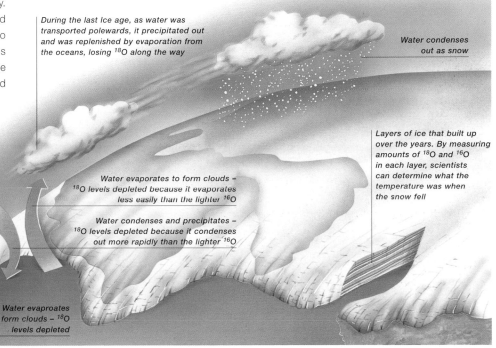

During the last Ice age, as water was transported polewards, it precipitated out and was replenished by evaporation from the oceans, losing ^{18}O along the way

Water condenses out as snow

Layers of ice that built up over the years. By measuring amounts of ^{18}O and ^{16}O in each layer, scientists can determine what the temperature was when the snow fell

Water evaporates to form clouds – ^{18}O levels depleted because it evaporates less easily than the lighter ^{16}O

Water condenses and precipitates – ^{18}O levels depleted because it condenses out more rapidly than the lighter ^{16}O

Water evaproates to form clouds – ^{18}O levels depleted

COPING WITH EXTREME COLD

THE HUMAN BODY IS NOT DESIGNED TO COPE WITH EXTREME COLD, AND YET THOUSANDS OF PEOPLE LIVE AND WORK IN TEMPERATURES WELL BELOW FREEZING.

The inhospitable weather in polar regions provides some of the best examples as to how people rise to the challenge of surviving in an extreme climate.

CREATING A MICROCLIMATE

The native peoples of the Arctic have evolved a lifestyle that enables them to combat the cold. By creating their own warm microclimate, through a combination of clothing and shelter, they have been able to maintain a subsistence economy – based on the indigenous wildlife – that has lasted at least 7000 years.

An important component of this way of life is effective clothing because, though Arctic peoples tolerate the cold better than those unused to it, body fat cannot protect them against temperatures well below freezing. The use of animal pelts and furs to produce clothes, mittens and boots, which keep the body snug even when temperatures fall to –60°C (–76°F), is therefore the real key to success.

Equally important are their shelters, built out of caribou or reindeer hides which, with the aid of only a small fire, can keep their occupants at a safe temperature, ie 0°C (32°F) or above, in even the bitterest cold. Another type of shelter, the igloo is made entirely of compacted snow and is a perfect design in terms of making effective use of available materials; it also prevents heat loss by producing the maximum living space with minimum surface area, thus reducing heat loss. Temperatures inside an igloo can rise to above freezing, though the frosty walls, by necessity, never go above 0°C (32°F).

THE ARCTIC DIET

The relevance of diet is the subject of ongoing scientific debate. Arctic peoples eat large quantities of animal fats and very few fresh vegetables. While providing plenty of energy to combat the cold, the fats are not regarded as the components of a healthy diet. The Inuit do not, however, have the same level of heart disease as people in industrial countries who also have a diet high in animal fats. The explanation probably lies in the high proportion of fish oils consumed by the Inuit, which reduces the risk of cardiovascular problems, while providing the fuel to survive the extreme cold.

ADAPTING TO THE COLD

Humans have a low tolerance to cold. Without protective clothing we will start generating extra heat and shivering at temperatures as high as 20°C (68°F), especially if the air is dry. However, over the course of generations there is some indication that humans can adapt slightly to cold conditions. Indeed, the people who are best equipped to handle the cold are not from polar regions but aboriginal Australians and Kalahari bushmen. These people seem to have adapted to the cold nights of desert regions as their extremities cool and their deep body temperature falls to levels that cannot be tolerated by the rest of us. So they can sleep comfortably in cold conditions that cause other people to start shivering, with smaller falls of both skin and body temperatures.

The best protection against cold is well-distributed body fat, and, in this respect, women are normally better protected. In fact, theoretically it pays to be short and round because heat production is proportional to body mass whereas heat loss increases with surface area. But, in practice, human beings do not appear to have the same capacity to adapt to cold as they do to hot climates. So although size and fat levels help a little, the only answer to cold climates is to create our own warm microclimate: regular cold baths are not the solution!

ADAPTATION VERSUS ACCLIMATIZATION

When held in freezing water the temperature of your hand or foot can do one of three things. It may swiftly drop to the water temperature and stay there, it may drop close to the water temperature and then cycle up and down, or it may adapt best to conditions by staying significantly warmer without cycling up and down at all. The type of response varies amongst individuals from different populations, suggesting that this is an adaptive response but it also depends on the amount of exposure the individual has had to cold. Men of black African origin show the least capacity to keep their extremities warm,

Below: An Inuk in Greenland dressed in white fox-fur clothes and holding a kakivak (an Inuit fish spear).

Above: An Inuit hunter about to enter his igloo in north-west Greenland at dusk. The internal lighting clearly shows how the igloo was constructed.

Europeans have an intermediate response, while the Inuit and people from high in the Peruvian Andes have the most effective response. But, there is evidence that people working in harsh conditions, such as deep-sea fishermen of Quebec, can acclimatize slightly. Repeated exposure builds up greater tolerance to such conditions because the blood vessels in the hands and feet develop the ability to dilate more effectively. Only the hands and feet are thought to be affected in this way.

GREENLAND COLONIZATION

The dying out of the Norse colony in Greenland is the only example of a well-established European society that was completely extinguished because of its inability to adapt to icy conditions. In 985, during a particularly warm period, an expedition led by Eric the Red from Iceland enabled some 300 to 400 colonists to set up two settlements on the western coast of Greenland (Osterbygd and Vesterbygd). By the beginning of the 12th century these communities had grown to include more than 300 farms with approximately 5000 inhabitants. They maintained herds of cattle, sheep and goats, exploited the plentiful wildlife of the region and received supplies of essential goods from Iceland and Scandinavia.

During the 12th century the climate in Greenland cooled sharply. Although it improved slightly in the 13th century, it became colder still during the 14th century. In addition, the North Atlantic became more stormy and more extensive pack-ice developed around Greenland. As a result, visits from Iceland became less and less frequent, and the last recorded contact was in 1410.

Archaeological studies of the settlements have revealed evidence of rising permafrost as the climate cooled, and the graves of settlers present a harrowing picture of malnutrition and debilitating diseases, which include rickets and tuberculosis. The debate continues as to whether other factors, such as hostilities from the Inuit, the plague or marauding pirates associated with the expanding whaling industry in the 14th and 15th centuries may have hastened the end of the colony in Greenland. There is little doubt that the deterioration in the climate was the principal cause for the demise, but failure to draw on the experience of the Inuit in adapting to the harsher conditions certainly contributed to the Norse downfall.

Above: The ruins of the Viking settlement at Brattahlid near Cape Farewell in Greenland.

Whaling records 62

Permafrost 78–9

Human response to heat 160–1

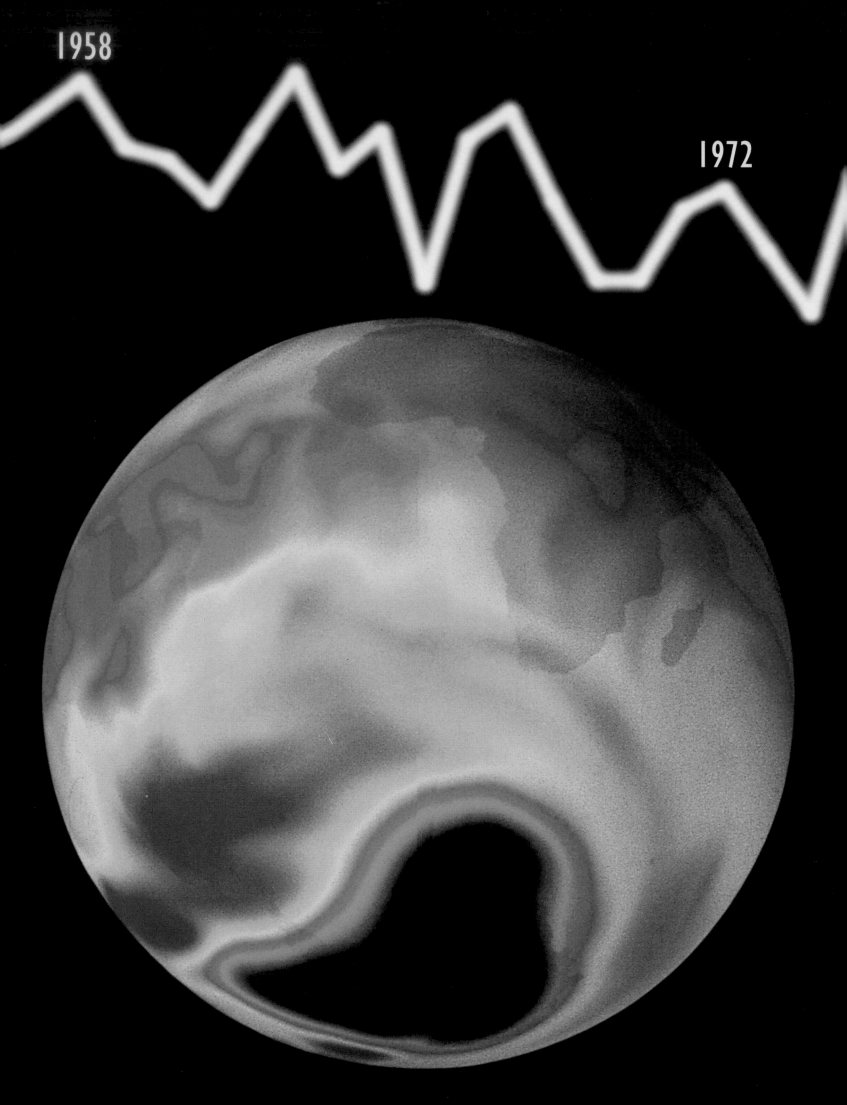

THE OZONE HOLE

The ozone hole is the most dramatic example of how human activities can affect the atmosphere, and has had a profound effect on current thinking about the damage that man is capable of inflicting on the climate.

The appearance of the ozone hole over Antarctica during the 1980s has led to the most concerted international action to cut down the emission of chlorofluorocarbons (CFCs) into the atmosphere. The success of these efforts provides a model for how we should handle other global environmental issues.

WHAT IS THE OZONE LAYER?

Ozone is a colourless gas and is essential because it protects us from the Sun's harmful ultraviolet rays, which are not absorbed by other atmospheric gases and so would otherwise reach the ground. It is present through the whole atmosphere but is concentrated in the stratosphere, 12–50km (7–30 miles) above the Earth. The ozone layer is formed by short wavelength ultraviolet radiation from the Sun, which breaks up some of the oxygen molecules in the upper atmosphere. These atomic oxygen fragments recombine with other oxygen molecules to form ozone molecules, each of which consists of three oxygen atoms. Ozone is reactive and in the presence of certain other chemicals, such as oxides of nitrogen and chlorine-containing compounds in the atmosphere, it can combine with them and revert back to oxygen.

The reason the ozone layer is so vulnerable is that there is so little of it. The total amount of ozone between us and space is equivalent to about one-third of a centimetre (one-eighth of an inch) of the gas at atmospheric pressure – no thicker than a slice of cucumber. Most ozone is produced over the tropics where solar radiation is strongest and most direct, and it is transported around the Earth by stratospheric winds. The amount of ozone in any particular place around the globe depends not only on this transport process but also on solar activity and sulphuric acid droplets which are emitted by major volcanoes. Both alter the rate at which oxygen is converted into ozone and vice versa.

Some atmospheric scientists and environmentalists were expressing fears as far back as the early 1970s that various human activities – such as the emissions from high-flying aircraft, the increased use of fertilizers and, most of all, the release of chlorofluorocarbons (CFCs) into the atmosphere – would damage the ozone layer.

THE ANTARCTIC OZONE HOLE

It was not until the mid-1980s that the environmentalists' worst fears were confirmed: the stratospheric ozone layer over Antarctica was disappearing each October. This depletion has taken place every October since then,

with lower levels of ozone being recorded almost each year. By 1994 the total amount of ozone in the continent's atmosphere had fallen by 70 per cent from the 1958 level, while in the stratosphere the destruction was almost complete. The record size of the hole over Antarctica was measured on 19 September 1998 at a huge 27.3 million km² (10.5 million square miles).

Scientific investigations of the chemical processes at work in the stratosphere over Antarctica confirmed that the build-up of CFCs in the atmosphere was the cause. These compounds interfere with the normal process of ozone formation by breaking down in the upper atmosphere to form highly reactive chlorine compounds, which, under some circumstances, can destroy ozone. This takes place at the end of each winter in the southern hemisphere. Every October an intensely cold vortex forms in the atmosphere over Antarctica. As the Sun returns, the combination of sunlight, chlorine compounds and ice clouds creates an ideal mix for destroying ozone.

THE REST OF THE WORLD

Changes in ozone levels over other parts of the world are much more difficult to predict. Scientists have discovered that the stratospheric ozone around the globe has declined by several per cent in recent years. This may be linked to the ozone hole over Antarctica, but solar activity and the eruption of Mount Pinatubo in 1991 have also been contributing factors.

There is increasing evidence that an ozone hole is now also starting to form over the Arctic at the end of each winter. This is less pronounced because the temperatures in the stratosphere over the North Pole do not fall as low as those over the Antarctic continent. In addition, the vortex is not as well-defined so more ozone can be transported from low latitudes to replenish any that is destroyed when the Sun returns to the Arctic at the start of spring. Nevertheless, the gradual development of this annual decline during the 1990s is a worrying trend. It shows that, unless the international accord to reduce the emission of ozone-depleting chemicals is sustained, the threat to the ozone layer will remain a major environmental issue.

Near ground level the problems with ozone are different. In cities during the summer, a cocktail of pollutants, mainly from motor vehicles, is cooked up in sunshine to form photochemical smog, an important ingredient of which is ozone. As a result, the amount of ozone in the lower atmosphere has been rising steadily in heavily populated areas of the world. Although this ozone helps to screen out ultraviolet rays, which were until recently absorbed by ozone in the stratosphere, it also damages vegetation and is harmful to humans even at low levels.

The scale and suddenness of the decline in the ozone layer shocked the scientific world, and led to the Montreal Protocol in 1987, which set about trying to eliminate certain CFCs from industrial production. It was subsequently revised in 1990 and 1992. As a result of this rapid action the global consumption of the most active gases fell by 40 per cent within five years and the levels of certain chlorine-containing chemicals in the atmosphere have started to decline. But it will be many decades before the CFCs already in the atmosphere are destroyed. In the meantime, there will be virtually no ozone in the stratosphere over Antarctica every October. Furthermore, if there is any backsliding in the international resolve to cut emissions of CFCs, any progress will be delayed yet further.

Left: Colour-coded satellite image showing how much ozone there was in the atmosphere in October 1995: yellow indicates the highest concentration of ozone and black, the lowest levels, which occur over Antarctica.

This graph shows the decline in ozone levels over Antarctica since 1958 – a fall of 70 per cent. Measurements were taken in October of each year.

1994

SEA ICE

A LAYER OF ICE FORMS ON THE SURFACE OF THE SEA
WHEN THE TEMPERATURE OF THE WATER FALLS TO
APPROXIMATELY –1.9°C (28.6°F).

Above: A view across the vast expanses of sea ice and icebergs in the Antarctic.

WHALING RECORDS
Analysis of logbooks submitted to the International Whaling Commission between 1931 and 1987, provides an example of how records of fishing can be used to draw inferences about climatic change. These data contain about 1.5 million catch records, which provide an indication of the ice extent because whales were most likely to be harpooned close to the edge of continuous sea ice where the highest density of krill is found. They indicate that the rough latitude of the ice-edge around Antarctica remained roughly constant at about 61.5°S between 1931 and 1955 and then shifted 2.8° southwards to around 64.3°S in the early 1970s, where it has remained since. This is a decline of some 25 per cent in the ice extent, which far exceeds any increase in the area since then. While this provides clues about long-term trends, it is hard to be certain that they have not been influenced by other factors not related to the climate.

The amount of sea ice varies from year to year but its average depth is about 3m (10ft). It can be attached to the land (fast ice) or float separately according to ocean currents and winds (pack-ice). The ice forms in the northern hemisphere and around Antarctica and has a major influence on the climate of much of the globe, as well as being a sensitive indicator of the effects of global warming.

ANNUAL VARIATIONS
In the northern hemisphere, the extent of pack-ice varies from a minimum of some 9 million km² (3.5 million square miles) in late summer, when it is confined to the central part of the Arctic basin, to a maximum of 16 million km² (6 million square miles) in late spring. At its greatest extent it fills most of the Arctic Ocean, Hudson Bay and the Sea of Okhotsk. It also spills out into the Bering Sea, but it is in the North Atlantic that it varies the most and therefore has the greatest influence. It drifts down the eastern coast of Greenland around Cape Farewell and up towards the Davis Strait where it joins a current down the coast of Labrador. Overall the average annual extent of the ice in the Arctic is some 13 million km² (5 million square miles) – nearly 20 per cent is made up of expanses of open water that become trapped in the pack-ice.

In Antarctica the pattern is simpler. Measuring some 12 million km² (4.5 million square miles), the sea ice covers less area than in the Arctic but it expands and contracts more. This gigantic annual pulse varies from a minimum of 3–4 million km² (1.2–1.5 million square miles) in February to a maximum of nearly 20 million km² (8 million square miles) in August. Within this ice boundary are some 3 million km² (1.2 million square miles) of open water known as 'leads' and larger areas of open water known as 'polynyas'.

DOES SEA ICE INFLUENCE THE CLIMATE?
The expansion and contraction of sea ice has major consequences for the climate in adjacent lower latitudes. The sea ice largely cuts off the transfer of heat and moisture to the atmosphere because it acts as an insulating layer. This means depressions at high latitudes pick up less energy and so the position of the sea ice influences their course. For this reason, the tracks of these storms tend to run at a latitude several degrees lower than the edge of the ice. Strong storms, in turn, can significantly alter the position of the ice and so the relationship between the sea ice and weather systems is interactive.

Long-term changes in the extent of sea ice in the North Atlantic have been linked with fluctuations in Europe's climate in recent centuries. Cold periods, such as the 1690s, occurred when the pack-ice to the north of Iceland was very extensive. This is thought to have increased the severity of cold northerly outbursts in the area.

RECENT TRENDS IN ICE COVER
From year to year, the amount of sea ice in the northern hemisphere is closely linked to the strength of the westerly winds during the winter months, especially in the Atlantic region. When there is a strong westerly flow of air, low-pressure systems regularly push through the Norwegian Sea and into the Barents Sea and keep the pack-ice back. At the same time this stream of depressions continually pulls Arctic air down the Davis Strait and pushes pack-ice further out beyond Newfoundland. During the late 1980s and early 1990s when the North Atlantic was particularly stormy, the ice conditions were exceptionally heavy in the West Greenland and Labrador region. This caused substantial problems for shipping in the Gulf of St Lawrence and along the coast of Labrador during the winter months between 1991 and 1994. It also produced consistently low sea-surface temperatures, which contributed to the disastrous decline in cod stocks in the Canadian Atlantic Ocean.

Above: Sunlight reflected by newly formed sea ice.

Below: A composite map of the northern hemisphere, showing observations made by the microwave radiometer on the Nimbus 7 satellite in February 1983. The extent of sea ice can be seen in white, permanent ice sheets in purple and snow cover in various shades of blue.

THE IMPACT OF GLOBAL WARMING

Modern technology has given us a clear idea of how the extent of sea ice varies. Satellite measurements of sea ice during the period November 1978 to December 1996 show that in the northern hemisphere the extent has decreased by a rate of 2.9 per cent each decade, while around Antarctica it has increased at a rate of 1.3 per cent each decade. Although these figures may seem alarming, there have been far greater shifts in the amount of sea ice in the past. Estimates of these early variations depend on sporadic observations from occasional scientific expeditions and the interpretation of other data such as the records of the whaling industry (see box far left).

NAVIGATION IN THE ARCTIC

Ships and boats in arctic regions must find a passage through the sea ice. In Russia there are routes along the north coast of Siberia from Murmansk to the Bering Strait, but they are only open for three or four months a year. Proposals to open up the Northwest Passage through the Canadian Archipelago ran into great objections in the 1970s. In 1969, the *Manhattan* completed its difficult journey through the Northwest Passage as a feasibility study by an American petroleum company to see if crude oil could be transported by tanker from Alaska to the east coast of the USA. As a reaction to this, the Canadian Government brought in legislation in April 1970 to extend its territorial waters and enacted the Arctic Waters Pollution Programme. Although the American Government has refused to recognize either of these acts, it did sign an Arctic co-operation agreement with Canada in 1988. So the combination of environmental objections, practical difficulties and diplomatic developments mean that the Northwest Passage is unlikely to be opened up as a navigable route in the foreseeable future.

ICE CORES

LOCKED UP IN THE ICE SHEETS OF ANTARCTICA AND GREENLAND CAN BE FOUND A WEALTH OF INFORMATION ABOUT PAST CLIMATIC CHANGE.

The permanent snows that compress down every year to form the ice sheets of Antarctica and Greenland not only contain a detailed record of the climate but also of other environmental changes extending as far back as 400,000 years. By drilling down into the ice sheets, scientists can extract ice cores and examine the different layers that have built up over the years.

DRILLING ICE CORES

Scientists have been working at altitudes of more than 3km (10,000ft) in the icy northern hemisphere polar regions for the last three decades to obtain a series of ice cores, some of which are more than 2km (6500ft) deep. These cores are cut using a special drilling bit and kept refrigerated, so that scientists can cut horizontal slices from any level and measure the properties of the snow when it fell long ago.

Right: Samples of snow being collected with a hand held corer.

IDENTIFYING ANNUAL LAYERS

In Greenland, snow accumulates quickly and so distinct annual layers have formed over the years. This enables scientists to estimate the annual temperature cycle at any time by analyzing the ratio of two isotopes of oxygen present in the ice (^{16}O and ^{18}O). In recent studies, they have identified layers going back some 15,000 years and were able to measure variations in snowfall over this period. However, it is difficult to identify every single annual layer and so this method lacks the precision of tree-ring analysis.

In Antarctica, snow accumulates much more slowly and so the layers are too thin to measure precise annual cycles further back than several hundred years.

LOOKING BACK STILL FURTHER

Scientists can date much older layers: although these do not contain clear evidence of annual variations, they do provide an insight into general trends in climate change. In Greenland, the ice near the centre of the ice sheet has dates back over 100,000 years, while beneath East Antarctica it is at least 400,000 years old.

In order to date this ice, scientists have to make assumptions about the direction and speed of its flow, particularly as the core approaches the bedrock, where there is much more risk of the layers being distorted by the shape of the underlying structure.

WHAT CAN THE ICE TELL US?

The dust content of the ice provides a measure of atmospheric circulation at the time the snow fell. Strong winds in mid-latitudes increase the amount of continental dust that is stirred up and transported to high latitudes. The amount of dust fluctuates sharply with shifts in temperature, so variations in oxygen isotope ratios can also be linked with atmospheric circulation patterns.

In addition, ice cores have made a major contribution to the study of the climatic effects of volcanoes because, by measuring the acidity of the different layers of ice, scientists are able to detect when major eruptions took place. The most important volcanoes in terms of climate are those that inject large quantities of sulphur compounds into the stratosphere, forming long-lasting sulphuric acid droplets, which may fall in rain or snow. The ice's acidity, therefore, is a particularly good measure of the climatic impact of past eruptions. Volcanoes that emit large quantities of sulphur into the atmosphere can have a cooling effect on the climate. In fact, scientists have found close correlations between volcanic eruptions, as revealed in ice cores, and cooling trends in summer temperatures in the northern hemisphere, as revealed in other proxy records, such as tree rings.

Analysis of the air bubbles trapped in the ice reveals the amount of each trace gas that was present in the atmosphere when the snow fell. Past fluctuations of trace gases such as carbon dioxide and methane are vital because they play a key role in the greenhouse effect and may have contributed to past climate change.

SUDDEN CHANGES

Some of the most interesting results to be obtained from ice cores relate to rapid changes in the Earth's climate. While the record for the last 10,000 years shows a remarkably stable climate, prior to this a much more dramatic and changeable picture emerges. When the climate was climbing out of the last Ice age, which lasted about 100,000 years and ended around 12,000–14,000 years ago, it went through enormous fluctuations in temperature and snowfall over a period of just a few years.

The rapidity of these changes came as a surprise to many climatologists and raised serious questions about the instability of the climate. Even more surprising is that one Greenland ice core showed how these sudden changes were a common feature of the last warm interglacial period, approximately 115,000–135,000 years ago. However, these results remain the subject of controversy between scientists because they may be due to distortions in the ice near the bedrock. More data from new cores are needed in order to resolve this issue.

Above: Samples of an Antarctic ice core being handled in a cold room. Slices of this core will be crushed and analyzed for both its dust and gas content and for the concentrations of its various constituent atoms, to identify features of the climate when the ice was formed.

THE VOSTOK ICE CORE

An ice core drilled by the Russian–French team at the Vostok research station provides climatic information dating back some 250,000 years. This record shows how the temperature has changed during the last two glacial periods and the warm interglacial about 130,000 years ago.

Temperature change (°C) vs Years ago (in thousands)

Dust and airborne particulates 166–7

TUNDRA AND TAIGA

CLIMATE OF TUNDRA AND TAIGA

THE COLDEST PLACES IN THE WORLD AFTER THE
POLES ARE THE UNDULATING TREELESS PLAINS OF THE
TUNDRA AND THE NORTHERN FORESTS OF THE TAIGA.

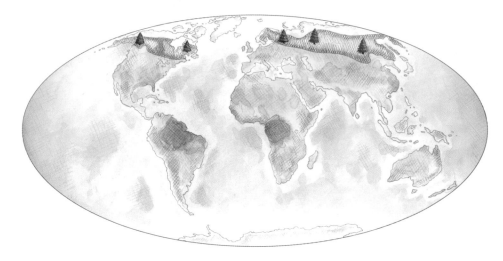

VERKHOYANSK, RUSSIA

■ *Average monthly temperature*

╱ *Monthly precipitation*

12 Outgoing energy

58–9 Coping with intense cold

The tundra and taiga are very different in terms of landscape and location. Tundra is the Finnish name for the treeless plains that stretch from northern Scandinavia to the Bering Sea and across Alaska and northern Canada. South of the tundra is the taiga, the great boreal forests, which spread from southern Scandinavia, across northern Asia to Alaska and Canada. These two regions share a similar climate – one of extremes, with freezing Arctic winters and mild warm summers.

Above: Tundra and taiga regions situated in eastern Siberia: the landscape of this area consists of grassy tussocks in the foreground and sparse trees behind.

ARCTIC WINTERS

Winter in the tundra and taiga is bitterly cold and poses challenges for both human and plant life. The coldest places are Verkhoyansk and Oimyakon, in eastern Siberia, where the average January temperature is an icy –50°C (–58°F). In North America it is not as extreme, dipping down to a mere –35°C (–31°F) on the northern coasts of Canada. Dense cold air sinks, hence the lowest temperatures are recorded in a thin layer of air just above the surface. In winter, very little sunlight reaches the tundra and much more than this is radiated out to space. The near-permanent snow cover reinforces these processes.

The only way this lost energy can be replaced is by weather systems transporting it in from warmer parts of the world. In Siberia, air masses from the Atlantic import large quantities of heat as far as the Urals, but run out of steam beyond that. This means that winter temperatures drop as you move eastwards. In addition, a large region of high pressure forms over eastern Siberia, producing calm weather and cloudless skies – perfect conditions for heat loss to space and maximum cooling. Low-lying regions in this part of the world have the coldest temperatures. In North America, the intrusion of warm air masses is largely blocked by the Rocky Mountains and other coastal ranges, so the coldest conditions are just east of the mountains.

WARMER SUMMERS

In mid-summer, there are long hours of sunlight and the snow melts away, which means that, unlike during the winter, large amounts of solar radiation are absorbed. This leads to surprisingly high temperatures, which do not vary very much, even over wide areas. For instance, Verkhoyansk has an average July temperature of 13.6°C (56°F), and has recorded highs of over 35°C (95°F). Only in regions close to the Arctic Ocean are the July figures constantly well below 10°C (50°F), and these areas experience conditions similar to those of mild winters in mid-latitudes.

Another unique feature of the tundra and taiga is that the annual temperature cycle follows the march of solar elevation: that is to say temperatures are dependent on how high the Sun is in the sky. This is because large expanses of the tundra and taiga are situated inland away from the sea. Oceans absorb more radiation than land masses and so maintain higher temperatures even after the Sun has passed its highest point in the sky. In maritime regions, therefore, the highest temperatures occur a month or more after the summer solstice. In the landlocked tundra and taiga, however, the peak is in early July and by August the temperatures start to fall sharply.

In spite of these extremely moderate summer temperatures, the frost-free season in these regions is never very long in duration. In the taiga, it varies according to latitude and distance from the sea and is never longer than about 50 days. Inland it lasts for an average of 50–90 days, but there is always a risk of a summer frost, so any crops that are grown in this region need to be hardy. In tundra regions frosts may occur in any month.

THE COLD DESERTS

The tundra is often referred to as a 'cold desert' because the precipitation levels are equivalent to a mere 25cm (10in) of rainfall a year. Much of it falls as rain in the summer and, while winter snowfall amounts can reach 1.5–2m (5–6½ft), there is no more than about 65cm (25in) in northern areas.

In the taiga, precipitation levels are slightly higher but still less than about 50cm (20in) per year. Most falls as rain in the summer, but there is some snow in the winter during the fierce storms that occur on the southern edge of the taiga in both Siberia and Canada when depressions move eastwards in the westerly winds. These bring driving snow and low temperatures as cold fronts caused by the depressions pass through the region. In Siberia these conditions are known as the buran, whereas in North America they are better known as a blizzard. This snow can accumulate for several months and cause spring floods. Summer rainfall is often inadequate to compensate for the evaporation and transpiration from plants during the long days of summer, and so the soils turn to dust and irrigation can be necessary in the few areas where crops are raised.

PERMAFROST – EARTH'S DEEP FREEZE

In most places, however, the soil never dries out in spite of low rainfall because of the permafrost – ground that is constantly frozen, sometimes to great depths. This can cause flooding in the snow-fed northward-flowing rivers in northern Canada and Siberia that freeze over in winter. Melting begins in the headwaters and middle courses of the rivers earlier in the spring than in the lower flood plains. The downstream channels are choked with ice and the ground around, which normally absorbs water, is frozen solid. This causes extensive flooding until the thaw at lower levels is well underway. During the summer, on the other hand, stagnant pools are a permanent feature of the landscape because permafrost permits little or no infiltration of water.

Above: Reindeer around a winter camp of the Saami people in Yamal, Siberia. The tents are made of reindeer skins.

A CLIMATE OF EXTREMES

TEMPERATURES IN THE TUNDRA AND TAIGA CAN VARY HUGELY FROM YEAR TO YEAR, FROM WINTER TO SUMMER AND EVEN FROM DAY TO DAY.

The extremes of temperature experienced in the northern plains and forests provide stern challenges for anyone or anything living there.

HIGHEST AND LOWEST

Some parts of eastern Siberia experience a temperature range of over 100°C (180°F), ranging from over 35°C (95°F) to –65°C (–85°F). The lowest figure ever recorded at Oimyakon is –71°C (–96°F), and an unofficial figure of –78°C (–108°F) has been claimed for this site. In North America the extremes are not quite as great: the lowest temperature ever recorded was –63°C (–81°F) at Snag, Yukon Territory, on 3 February 1947, while the highest summer temperatures rise above 30°C (86°F). The highs are therefore similar to those recorded further south and not far below those in many parts of the tropics but the lows are considerably colder than any that are ever experienced in temperate regions.

WHY IS THE CLIMATE SO VARIABLE?

Conditions across northern Canada and Siberia vary dramatically with changing circulation patterns because they are so dependent on whether air is transported into these regions either from the cold Arctic Ocean or from warmer areas. Sometimes weather systems become trapped and air from further afield cannot be transported as far as the tundra. When this happens the air becomes stagnant and extremely cold. At other times of the year, weather systems sweep through the area, bringing cold and warm air with them, which leads to ups and downs in temperature. As a result, the tundra and taiga regions have greater temperature variations from year to year than anywhere else in the world, with Verkhoyansk experiencing temperature differences of over 20°C (36°F) between the coldest and warmest Januarys of the 20th century, and 10°C (18°F) in the July figures.

INTERPRETING EXTREME VARIABILITY

The extreme variability of weather in the tundra is significant because it plays an important part in meteorologists' interpretations of the current climatic warming trend in the northern hemisphere. There has been an increase in winter temperatures over land, especially in Siberia, which has been caused by a speeding up of atmospheric circulation in the northern hemisphere – this results in more warm air being transported deep into the region. Because the fluctuations of temperature from year to year in the tundra are so dependent on the strength of this atmospheric circulation, it is important to understand what causes changes in the patterns of air flow around the planet. We are not sure as yet whether circulation changes are caused by human activities or whether they are just a feature of the natural variability of the Earth's climate. So, although the variations in surface temperature are real, they must be considered in the context of the overall behaviour of the atmosphere. Only then can meteorologists assess the true scale and permanence of any changing temperature trends.

BREEDING GROUND FOR COLD WAVES

The frigid air that forms at high altitudes in the middle of continents during the winter months is the source of one of the most dramatic weather phenomena of the east-coast regions of Asia and North America: cold waves. These intensely cold air masses sweep southwards bringing with them sudden drops in temperature to low latitudes – the Canadian Arctic air, for example, often reaches the coast

Right: In summer the tundra is a carpet of wildflowers. This picture was taken at Akkani, Chokotka, in Siberia.

15 The atmospheric motion

58–9 Coping with extreme cold

Left: The frozen mountains situated in eastern Siberia. In the valleys of these isolated peaks, the lowest temperatures in the northern hemisphere are recorded in winter.

of the Gulf of Mexico; while temperatures can fall as low as –15°C (5°F) from Galveston to Tallahassee, and have been known to drop by more than 20°C (36°F) in just a few hours. The most extraordinary example of this type of climatic event took place on 23 January 1916 at Browning, Montana, when the temperature fell from 7°C (44°F) to –49°C (–56°F) in just 24 hours.

The cold waves are relatively shallow surface layers of air extending to a height of approximately 1000–2000m (3300–6600ft). As they swoop south-eastwards they thin and diverge, warming only slightly as they pass over land masses, but rapidly warming when they reach the ocean. In North America the course of cold waves is strongly influenced by the Rocky Mountains, and each winter there are, on average, three or four of these icy blasts. Across North China and Korea they occur approximately once a week. As they sweep out across the Sea of Japan they swiftly warm and pick up copious amounts of moisture, producing heavy snowfall on the western slopes of the Japanese Alps.

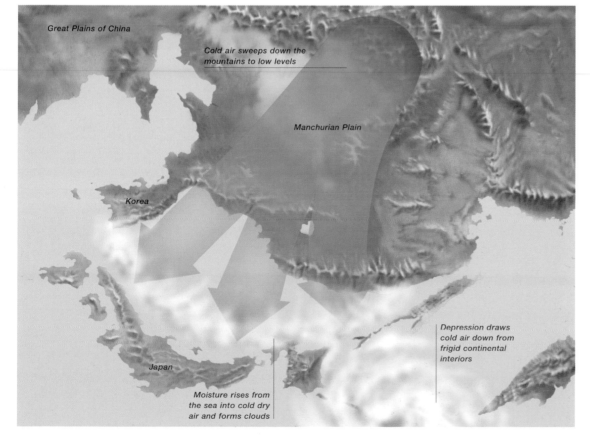

Great Plains of China

Cold air sweeps down the mountains to low levels

Manchurian Plain

Korea

Japan

Moisture rises from the sea into cold dry air and forms clouds

Depression draws cold air down from frigid continental interiors

COLD WAVES

When depressions move along the eastern coast of the northern continents in winter, they sometimes draw blasts of intensely cold air down from inland areas. Here a cold wave sweeps out of Siberia over Korea towards Japan.

TUNDRA FLORA AND FAUNA

ONLY THE MOST HARDY AND ADAPTABLE FLORA AND
FAUNA MANAGE TO SURVIVE IN THE HARSH AND
VARIABLE CLIMATE OF THE TUNDRA.

WITHSTANDING THE COLD
When low temperatures cause a plant to freeze, ice forms in between its cells – extracellular freezing. This process removes water from inside the cells and so protects them from destructive intracellular freezing – the formation of ice within the cell. This dehydration also helps as it increases the concentration of solutes in the cell fluid, which protect it but can also cause damage. Some species are able to exploit the fact that water can be supercooled as low as –40°C (–40°F) without freezing. This figure coincides with the lowest temperature limit that many plants can tolerate, and so supercooling may be the main means of avoiding intracellular freezing. Certain species of conifer can tolerate temperatures as low as –70°C (–94°F) in their buds, and sometimes in their leaves, suggesting that the presence of solutes in their cells enables supercooling to occur down to these low temperatures. Tundra lichens never freeze and have the greatest tolerance to the cold. They are also able to withstand rapid and extreme temperature changes.

The way in which flora and fauna survive in the bitter and variable tundra conditions provides an insight into the impact of the climate on many aspects of life.

TUNDRA FLORA

Tundra flora comes in four principal forms: moss-tundra, lichen-tundra, dwarf-heath tundra and arctic-alpine tundra. These different forms often grow in close vicinity to each other because they are all dependent on local factors such as variations in soil, soil-moisture, duration of snow cover and the amount of shelter available.

Moss-tundra grows in damp areas, often forming peaty mounds, interspersed by puddles of water. Lichen-tundra grows on drier and shallower soils, where thin layers of peat accumulate. Dwarf shrubs and herbaceous plants such as grasses and rushes are often found growing between the lichen masses. Dwarf-heath tundra is made up of stunted or creeping, slow-growing shrublets, mostly of the heath family, usually no more than 20cm (8in) high. These are mostly found on sandy well-drained soils that dry out in summer and are low in nutrients. Arctic-alpine tundra is found on higher ground but can spread to lower latitudes.

Below: Dwarf willows growing in mountains in Iceland. This species is a good example of the hardy trees and shrubs that can withstand the intense cold of the tundra.

TUNDRA MEETS TAIGA

On the southern edge of the tundra, where it merges into the taiga, there is increasing diversity of plant species and the landscape is made up of a mosaic of trees and open spaces that is often defined by topography and by the quality of drainage that is available. The northern limit of the taiga is usually defined as coinciding with the 10°C (50°F) isotherm (a line on a map that links places of equal average temperature) in the warmest month of the year. In reality the marginal zone is a lengthy transition between forest that is suitable for economic exploitation, through ever thinner and more vulnerable stands of trees past the poleward limit of woody perennial trees (which are at least 2m/6½ft in height with a single stem), to the limit of other hardier species of tree.

SURVIVING THE WINTER

Some species manage to survive the winter in spite of the intensely cold temperatures, frozen soil and the problem of accelerated evaporation and transpiration brought on by the low relative humidity and high winds. The evolution of dwarf trees and shrubs, and of low-lying plant forms, is not simply a matter of being stunted by these harsh climatic conditions, but shows how the plants exploit snow cover in order to protect themselves against cold and dry conditions. The same approach is adopted by plants that grow in low hemispherical hummock shapes or in pillow forms, with their leaves protecting them from the wind. These physical forms, combined with drought tolerance, ensure extreme winter hardiness.

THE ARRIVAL OF SUMMER

The short summer on the tundra brings a complete transformation. The flora and fauna burst into activity to exploit the long hours of sunshine and the ground is covered with a mass of flowers. The thaw produces shallow pools of water, which warm in the sun and encourage vast swarms of insects. These provide fodder for many local species and for the flocks of migratory birds – notably ducks, swan, geese and waders – which fly in from the south to feast on the plentiful food supply.

POPULATION DYNAMICS

This sudden abundance of food combined with variations in the weather from year to year produces dramatic changes in the numbers of certain animals, and there have been many attempts to establish a link between the two.

The best-known relationship is that between the lynx and its main prey, the snowshoe hare. Evidence from fur company records in northern Canada between 1821 and 1934 shows that the lynx population fluctuates every 11 years as a result of fluctuations in the number of hares. Some naturalists have tried unsuccessfully to link this cycle with the sunspot cycle, which is of similar length and may cause variations in the climate.

Among larger animals, reindeer have the widest range of population numbers with seasonal migrations from summer tundra to winter taiga grazings. Their numbers appear to fluctuate roughly every 100 years. No explanation for this fluctuation has been identified and, as this species has been semi-domesticated in Eurasia by the Saami peoples, it hard to see how the climate can be the principal

cause. In North America, where the reindeer's close cousin, the caribou, has proved impossible to domesticate, more natural factors may be at work.

Among smaller animals, perhaps the best-known cycle of all is the one that influences the number of lemmings. These widely distributed, abundant little rodents have a marked population cycle of 3–6 years: population levels in small areas can rise above 200 per hectare (80 per acre) and then crash to less than one per hectare. Various suggestions attempt to explain this cycle: variations in winter temperatures, snow cover, nutrition, the number of predators and hormone levels of the lemmings, or any combination of these factors. But there is no established Arctic weather cycle that explains this puzzle of nature.

Above: The Canadian lynx (top) and the snowshoe hare (above). The predator–prey relationship between the two produces one of the best-known examples of linked population fluctuations in the animal kingdom.

THE NORTHERN FORESTS

FORESTS, WHICH ARE OFTEN DESCRIBED AS THE
LUNGS OF OUR PLANET, ARE A MAJOR COMPONENT
OF THE EARTH'S BIOSPHERE.

The great northern (boreal) forests stretch from Scandinavia across northern Asia to Alaska and Canada. They play an extremely important part in the Earth's biosphere because of the high levels of photosynthesis – the process by which plants absorb carbon dioxide and emit oxygen.

SPECIES OF THE TAIGA

The northern forests (usually known by the Russian word taiga, which means swamp forest) do not contain a wide variety of tree species and are made up principally of evergreen coniferous trees. Their needle-like leaves can withstand several months of freezing conditions and snow cover. They can also reduce water loss (transpiration) by closing up their pores, a process useful in cold or dry spells. The dominant species in western Eurasia are Norway spruce and Scots pine, while in northern Siberia 90 per cent of the forest is made up of larch. Some deciduous trees can also survive these harsh conditions: birch and aspen form sub-alpine woods at high levels, and alder and willow thrive in wet areas.

OTHER FEATURES OF THE TAIGA

In many areas mature taiga forms such a dense canopy that only lesser forms of vegetation can compete to a limited extent and tend to be no more than an understorey of low dwarf-heath shrubs or thick ground-cover plants. However, when an area of forest is cleared, often as the result of human activities, members of the heath family, such as bilberry, cowberry and heather, and related dwarf shrubs, can take over. Mires or bogs tend to cover large areas where drainage is poor.

On the northern edge of the forest, growth is slower and the trees become stunted and more widely spaced than elsewhere. Increased soil moisture leads to the build-up of peat on the forest floor – this locks up the vegetation and the soil nutrients, which reduces the capacity of the forest to recycle this material and slows growth. The extreme freeze/thaw cycles caused by permafrost make the ground heave. In spite of well-developed lateral root systems, the disturbance can dislodge and fell trees, producing what is called 'drunken forest'.

HISTORY OF THE TAIGA

In spite of their great extent and sense of permanence, the boreal forests are relatively young compared with many other parts of the biosphere. They moved southwards during the last Ice age, and then remigrated northwards and re-established when the climate became warmer again. Analysis of pollen records in North America show that spruce were the first trees to colonize the region during this period. The most northerly extent of the forests started

to grow around 5000–8000 years ago. The gradual cooling of the climate from then until the beginning of the 20th century resulted in the northern limits of the forest receding by about 100–200km (60–120 miles) in parts of Canada and Siberia.

IMPACT OF GLOBAL WARMING

Recent global warming seems to have caused tundra and taiga regions to spread northwards again. Although small local changes may have been caused by shifts in regional climatic patterns, satellite observations between 1981 and 1991 show that, on average, the biosphere between 45 and 70°N experienced more photosynthesis than usual, indicating an increase in the amount of vegetation. This was probably caused by the rise in spring temperatures, which lengthened the growing season at high latitudes. Plants draw more carbon dioxide out of the atmosphere in the summer than the winter and, although levels of the gas are increasing generally, there is a greater swing from the high winter figures in the Arctic atmosphere to the lower summer figures, suggesting that there are more trees.

However, these results have the obvious limitation that they only cover a short period. Moreover, they coincide with a period when snow cover in the northern hemisphere declined sharply and, as this fall has not been sustained during the 1990s, more observations are needed before any conclusions can be drawn about how rapidly the forests are going to spread northwards in the future.

The possible increase in the growth of the taiga touches on another interesting aspect of the current global warming, which relates to our limited knowledge of

Above: Snow-covered larch trees in winter in the northern forests of Siberia.

Above: The lush summer growth that takes place in the Siberian taiga is stimulated by the long days during the summer months. Plants have to complete their development before the intense cold returns in early autumn.

decreasing levels of carbon dioxide caused by natural factors. Measurements in the early 1990s suggested that the amount of carbon dioxide being absorbed at high latitudes was much greater than previously estimated, to the extent that half of all the emissions from burning fossil fuels were being mopped up by an unidentified source. This 'Great Northern Sink' has been the subject of a great deal of speculation among scientists. The obvious candidate is increased levels of photosynthesis, either by microscopic algae in the North Atlantic Ocean or by the vast expanses of the northern forests.

FOREST FIRES

The forests of the taiga are particularly vulnerable to fires set off by lightning because of the dominance of conifers, with their resinous wood and litter of needles and pine cones, and the limited number of other tree species. Forest fires are a natural part of the taiga ecology and occur roughly every 50–200 years. They often burn large areas of forests: in one extreme case in western Alaska over 400,000 hectares (1 million acres) were destroyed. The frequency of these conflagrations depends on climatic conditions and the nature of the forest. They are less frequent where the rainfall is higher and more frequent where drought is common. They also become rarer towards the northern limit of the taiga.

It is now accepted that forest fires are essential in renewing forest ecology, and have been part of their development throughout time. Without this regular recycling of nutrients, too much matter becomes locked up in dead material and there is a danger of mosses and peat bogs taking over. So, in spite of their apparently destructive impact, such fires are a vital component of the long-term survival of the forests.

SNOW COVER

SNOW-COVERED AREAS ARE EXTREMELY REFLECTIVE
AND SO ANY CHANGES IN SNOW EXTENT AFFECT
RADIATION LEVELS AND THUS THE CLIMATE.

Changes in the amount of snow cover from year to year can have a powerful impact on the climate. Large covered areas can increase any global cooling trend because snow reflects so much sunlight into space.

THE REFLECTIVITY OF SNOW

When viewed from space the most highly reflective surfaces of the Earth are the snow-covered polar regions. Fresh snow can reflect as much as 90 per cent of the sunlight it receives, but as it ages or starts to melt its albedo (reflectivity) will drop noticeably. The average albedo of snow-covered forests is much less than that for snowy treeless country, the former reflecting around 50 per cent of sunlight and the latter nearly 70 per cent.

CHANGES IN SNOW COVER

Any increase in the amount of snow cover can have a cooling effect on the climate – the longer the unusual amount of snow persists, the longer the climatic consequences will persist. On the other hand, any reduction in the amount of snow cover because of a rise in global temperatures will contribute to the warming trend because of the increased absorption of sunlight by the newly exposed soil. The potential for the runaway effect of this positive-feedback mechanism, due to fluctuations in snow cover, has been the subject of a great deal of debate by climatologists because such an effect could destabilize the climate.

GLOBAL EFFECTS OF SNOW COVER

One way of examining the possible positive-feedback effects of snow cover is to see whether the extremes in the most severe winters have a long-term climatic impact in parts of the northern hemisphere. This has been extended to see whether fluctuations in the extent of the snow cover in the northern hemisphere have similar effects.

To have a lasting climatic impact, changes in snow cover in the northern hemisphere must last for a significant period of time. The average extent of snow cover varies from over 45 million km^2 (17 million square miles) in mid-winter to a little over 4 million km^2 (1.5 million square miles) in August. The most extensive anomalies are likely to occur in the winter half of the year, but to have an significant climatic impact they must last well into the spring and summer when there is most sunlight at high latitudes. Satellite records show, however, that the greatest variations proportionally occur between different summers – in some years the cover is twice as extensive as in others. But, in terms of area covered by snow, the proportionally smaller changes from winter to winter represent much larger climatic variations. More importantly, abnormal snow cover is a transient phenomenon and large anomalies, at any time of the year, rarely last for more than a few months.

REGIONAL EFFECTS OF SNOW COVER

At a regional level, changes in snow cover may exert a greater influence than at a global level. The extensive snow cover across Europe during the exceptional winter of 1962–3 is thought to have been partially instrumental in sustaining the cold weather. This winter was the coldest since 1830 and meant that virtually all of Europe north of the Alps was covered in deep snow throughout January

Above: An aerial view of the snow-covered northern forest of Labrador, Canada, showing how the trees stand out darkly against the snowy surface.

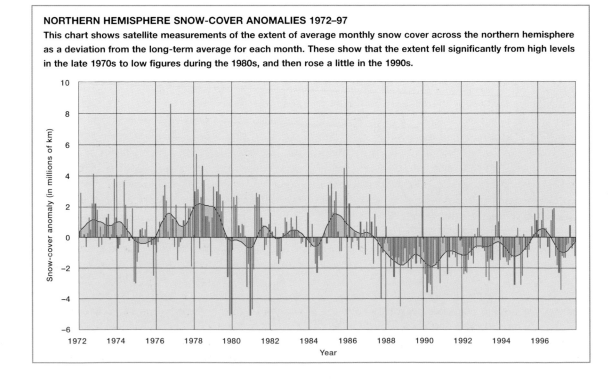

NORTHERN HEMISPHERE SNOW-COVER ANOMALIES 1972–97
This chart shows satellite measurements of the extent of average monthly snow cover across the northern hemisphere as a deviation from the long-term average for each month. These show that the extent fell significantly from high levels in the late 1970s to low figures during the 1980s, and then rose a little in the 1990s.

and February 1963. The prolonged snow cover helped to sustain the high-pressure region over Scandinavia, which was the main feature of the abnormal weather. Similarly, in the USA the extensive snow cover that built up during the record-breaking cold spell in December 1983 helped prolong the wintry weather. During January 1984 it is estimated that in parts of the Midwest the daytime maximum temperature was 5°C (9°F) lower than would have been expected on the basis of atmospheric conditions.

LONG-TERM CONSEQUENCES

In the long term, the satellite observations show that the extent of snow cover noticeably declined between the late 1970s and around 1990, since when it has, if anything, increased a little. The overall trend correlates closely with changes in the average temperature at high latitudes in the northern hemisphere. This suggests that fluctuations in snow cover are driven principally by long-term changes in global circulation patterns and, in particular, by the North Atlantic Oscillation (NAO). Therefore they will only contribute significantly to long-term shifts in the climate if they reinforce the impact of other large-scale factors. For example, an increase in snow cover seems to have

contributed to the onset of the ice ages because it reinforced the impact of the main cause: variations in the Earth's orbit (the Milankovitch effect).

All this evidence suggests that snow-cover fluctuations have a substantial short-term regional impact on temperatures during the winter half of the year but, in general, the long-term variability of the global climate overwhelms their temporary influence.

DEFORESTATION AND SNOW COVER

One aspect of snow cover that may have a more significant impact is the removal of the northern forests. Although most concern about deforestation relates to the tropics, its consequences further north may be far more significant. Snow-covered forest absorbs about half the sunlight falling on it, whereas snowy, open ground absorbs barely one-third of available energy. This means that the effect of deforestation would be to produce a dramatic cooling across the northern hemisphere, especially at the end of winter and early spring. Both computer models and practical developments in numerical weather forecasting suggest that, where large areas of forest are removed, temperatures could cool by as much as 10°C (18°F).

Above: The village of Radstadt in Austria, showing how the snow-covered fields surrounding the village reflect sunlight efficiently.

PERMAFROST

THE GROUND BENEATH MOST OF THE TUNDRA AND MUCH OF THE TAIGA IS FROZEN TO A GREAT DEPTH – RATHER LIKE NATURE'S ANSWER TO A DEEP FREEZE.

Above: An example of an ice lens in the permafrost close to the surface.

The frozen ground beneath the tundra and taiga controls many aspects of life in these regions. Use of the land for agriculture is virtually impossible and care has to be taken when constructing buildings. The Inuit, however, turn it to an advantage by using it as an all year round deep freeze for food products.

WHERE IS PERMAFROST FOUND?

The belt of northern hemisphere permafrost begins in eastern Europe and extends through Siberia, Alaska and Canada as far as Greenland. The area of continuous permafrost extends northwards from about the –5°C (23°F) isotherm of average annual surface air temperature. At its southern limit, there is a mean annual temperature of –1°C (30°F) and a critical snow depth of 40cm (16in). In northern areas it usually extends to a depth of 1500m (5000ft), and also out under the Arctic Ocean.

The depth at which permafrost occurs varies from around 20cm (8in) in the north to 1.5m (5ft) or more in the south (in the latter it becomes discontinuous from place to place, which means that it is not a permanent feature). It forms an impermeable barrier and it is only the top active layer that thaws out during the short summer period. Both the depth of the active layer and its duration of thaw vary from year to year. Permafrost impedes downward drainage, accelerates run-off of rainfall and causes saturated or waterlogged soils above. The active soil layer expands and contracts during the seasonal freeze/thaw cycle because it both thaws and then freezes from the top downwards. This process sets up considerable stresses in the soil, which cause the ground to move in different directions. The result can be seen in the shape of pingos (icy conical hummocks, up to 30m/100ft high) and polygonal patterns across the tundra (see below).

THE DANGER OF HUMAN ACTIVITIES

The fragile balance of both the tundra and taiga environments, and the sensitivity of the permafrost to disturbance, mean that the impact of human activities tends to be very high. When the vegetation is removed, the permafrost is far more likely to melt and cause flooding. Whether it be the extraction of oil and gas from the massive reserves in Alaska and Siberia, and the mining operations in these areas, or the economic exploitation of the forests for timber, great care is needed to minimize the damage. This is of greatest concern in the tundra, where vegetation takes such a long time to regenerate. Even before current pressures developed, expansion of reindeer herds had led to overgrazing. The natural sedge–lichen–grass vegetation has now mainly been replaced by grass. By the beginning of the 20th century, lichen pasture in northern Europe and Russia took up a mere 1 per cent of its original area.

MEETING THE CHALLENGE

Any developments in permafrost regions have to take account of the need not only to avoid melting the subsoil, but also to accommodate the movement of the soil, which places so many additional stresses on buildings. An enormous amount of effort, for example, went into the design of the Trans-Alaska pipeline to ensure that it would transmit as little heat as possible to the ground (see above right). In contrast, the Russian system was not as successful and caused extensive pollution in Siberia due to a rupturing of oil pipelines.

Perhaps the most significant climatic consequence of industrial activity in these regions is the development of what is known as 'arctic haze'. This layer of pollutants,

Right: An ancient pingo stands out from the surrounding permafrost and has typical polygonal forms on its surface.

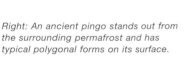

forms in late winter and spring across the Arctic Ocean and is trapped in the intensely cold layer of surface air. Climatologists have established that this haze alters the radiative properties of the atmosphere, but it is not clear what the climatic impact will be because so few measurements have been made of atmospheric conditions both before and after the haze appeared.

METHANE HYDRATE

The possibility of the permafrost melting as a result of global warming has serious implications because it contains large quantities of methane – a powerful greenhouse gas. This gas is present, locked up in ice crystals, in the form of methane hydrate. Compounds similar to this can be formed with a variety of gases and occur widely in nature under suitable conditions of temperature and pressure. To remain stable at temperatures typical of the Arctic, methane hydrate has to remain at high pressures such as those found at a depth of about 200m (650ft) or greater. In permafrost regions, it is stable at depths of about 250–1000m (800–3300ft).

It is estimated that there is the equivalent of two trillion tonnes of carbon locked up in methane hydrates frozen in the ground. This is three times the amount of carbon currently present in the atmosphere, and about a quarter of the global fossil-fuel reserves. Although most of this gas would not be released by any rise in temperature that is currently predicted, it needs to be considered in the long term if some of the more rapid warming scenarios for the Arctic begin to look realistic; this is because it represents a positive-feedback mechanism, which could reinforce any such rapid warming.

Above: The Trans-Alaska oil pipeline, which is suspended on pillars to insulate it from the permafrost below and hence prevent any melting of the fragile soil structure during the summer months.

 Continuous permafrost

 Discontinuous permafrost

 Sporadic permafrost

CROSS-SECTION OF PERMAFROST
The depth of permafrost in north Canada increases with latitude from a discontinous layer a few tens of metres thick at 60°N to several hundred metres at 70°N. The active layer, which melts in summer, becomes thinner as you go further north.

THE EXTENT OF PERMAFROST
The distribution of permafrost conditions in the Arctic, showing the zones of continuous, discontinuous and sporadic permafrost.

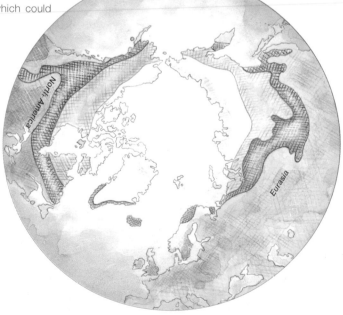

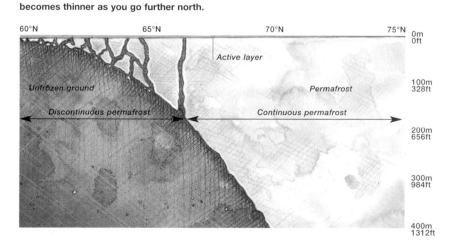

| 60°N | 65°N | 70°N | 75°N | 0m 0ft |

Active layer

Unfrozen ground

Permafrost — 100m 328ft

Discontinuous permafrost — Continuous permafrost — 200m 656ft

300m 984ft

400m 1312ft

Pressure systems 109

Global warming 184–5

MOUNTAIN REGIONS

CLIMATE OF THE MOUNTAINS

MOUNTAINS ARE UNIQUE BECAUSE THEY GENERATE
THEIR OWN WEATHER AND SO TRANSCEND OUR
USUAL INTERPRETATION OF CLIMATIC ZONES.

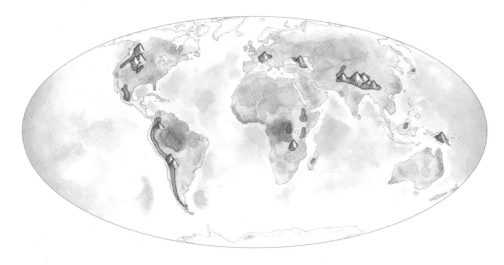

up to which snow melts away in summer. And they must have enough snow to produce distinctive glacial landforms. Often termed 'high mountains', peaks meeting these criteria vary in height according to their location, ranging from 1500m (5000ft) in the Alps to around 4500m (15,000ft) in the Andes on the equator.

THE BASICS OF MOUNTAIN WEATHER

The most basic feature of the mountain climate is that the higher it is, the colder it becomes. On average the temperature drops by 6.5°C for each kilometre (or 3.6°F every 1000ft) in altitude. This decline in temperature with altitude is known as the lapse rate and can vary according to weather conditions: moist warm air cools more slowly than cold dry air because it releases heat as water vapour condenses out. The lapse rate in tropical mountains is less rapid than that at higher latitudes.

Another obvious feature of the climate of mountains is that the amount of rain and snow increases with altitude. In mid-latitudes the amount of precipitation doubles between sea level and about 1500m (5000ft), and then doubles again by the time it reaches 3000m (10,000ft). The heaviest precipitation occurs on the windward side of mountain ranges, falling off sharply on the sheltered leeward side to create an area of low rainfall known as a rain shadow. This happens because, when weather systems pass over mountains, the air is forced to rise and, as it cools, rain and snow condense out. By the time the air reaches the leeward side of the mountain, therefore, there is little moisture remaining in the air.

Below: Hidden Peak in the Karakoram range in Pakistan. The mountain is seen from the south-west, near the base camp situated on the Abruzzi Glacier.

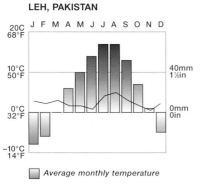

SONNBLICK, AUSTRIA

Millions of us visit mountainous regions every year to enjoy winter sports. However, mountain weather can be difficult to predict and there are a number of challenges to be faced: torrential rain and snow, ferociously strong winds, bitterly cold temperatures and a high risk of sunburn from the Sun's harmful ultraviolet rays.

WHAT ARE MOUNTAINS?

It may seem odd, but we must start by defining what we mean by mountains: peaks that are significantly higher than the surrounding countryside. They must have a clear upper tree-line, where forests thin out, and snow-line, the point

LEH, PAKISTAN

☐ *Average monthly temperature*

╱ *Monthly precipitation*

In addition, when air passes over mountains, it is squeezed by the flow process, which leads to an increase in wind speed over the crest of the range and then a decrease on the sheltered leeward side. Whilst moving over the higher peaks, the air funnels and accelerates through passes and over cols, and eddies and swirls down valleys, creating regional effects that lead to subtle variations in snowfall and winds, including pockets of high snowfall (referred to as Schneewinkel) and warm descending winds (known as Chinook or Föhn).

MOUNTAIN SNOW

Variations in temperature and snowfall, together with the aspect of the slope in question, determine how long snow lies and whether glaciers form. As far as skiers are concerned, a rule of thumb is that reliable snow cover occurs when the average monthly temperature is below –3°C (27°F). Places that are slightly warmer but have heavier snowfall build up a base that survives occasional thaws. The level of permanent snow cover (the firnline) is dependent on the temperature and on the amount of snowfall and sunshine.

The heaviest snowfall occurs on the windward slopes of mountain ranges in areas influenced by mid-latitude westerly winds, such as the Cascade Mountains in north-western USA and the Southern Alps of New Zealand. The highest figure of snowfall ever recorded was 31.1m (102ft) at the Paradise Ranger Station, on Mount Rainier, near Seattle, Washington, at an altitude of 1655m (5428ft) in 1971–2. By March 1972 the snow had compressed down to a depth of 9.3m (30ft).

The immense mounds of snow that accumulate on the windward sides of mountains form rivers of ice that can sometimes stretch all the way down to the sea, even in places where the climate is relatively mild, such as southern Alaska, southern Chile and New Zealand. In many other parts of the world, from the equatorial Andes and the Himalayas to Norway, Iceland and Spitzbergen, there is enough snow to maintain sizeable ice fields and glaciers. These icy regions are, however, under threat from the consequences of global warming.

SUNSHINE

Mountainous regions often get much more sunshine than the lowlands around them. For instance, during the winter, the uplands in the Alps get far more sunshine than areas further down the slopes because they rise above the low-level clouds. In the winter, temperatures in the valleys falll to lower values and the air close to the snow-covered slopes cools and slips downhill to the lowlands. This creates a temperature inversion, which traps pollutants and maintains fog and low cloud in the valleys. Conversely, in summer, much of the snow melts away and the lowlands receive more sunshine. This causes convection and warm air flows up the mountain to form clouds around the mountain tops.

Overall, the annual figures are not noticeably different but there are serious implications for recreational activities, especially skiing. The other major consequence of sunshine at higher altitudes is the increased level of ultraviolet radiation that is present, which can damage unprotected human skin.

Above: The Jungfrau (4158m/13,642ft), in the Bernese Oberland, Switzerland, rising up through the clouds. The mountain is seen here from the summit of the adjacent peak of the Mönch to the north-east.

AVALANCHES

Avalanches are awe-inspiring natural spectacles, but they can be immensely dangerous as they sweep down mountainsides destroying everything in their path.

Avalanches are a major hazard in many mountain regions. In the Alps alone, they cause some 100 deaths a year. They are a threat not only to people who live there but also to the millions of mountaineers, skiers and walkers who venture into snowy regions every year to enjoy winter sports. The biggest avalanches occur in the Andes, Himalayas and in Alaska. In 1962, 4 million tons of snow and ice cascaded off the peak of Nevado de Huascarán in Peru, sweeping away a town and seven villages and killing 3500 people. However, it is in the heavily exploited European Alps and the Rockies of North America that more and more avalanches are occurring.

Different types of avalanches

Avalanches fall into two broad categories: loose-snow avalanches and slab avalanches. A loose-snow avalanche can take place in both wet and dry snow. As it builds up, it gains in volume so a small slide of snow may eventually develop into a very large and dangerous avalanche. In contrast, a slab avalanche starts when a block of compacted snow, shaped rather like a paving stone, starts to move down the mountain. While this type of avalanche can unsettle the snow around it, its major impact and danger lie in the initial release of snow.

Both types of avalanche come in many forms and sizes. Loose-snow avalanches range from small slides of light powder, which only pose a threat to mountaineers on the steepest slopes, to huge affairs that result in tens of thousands of tons of new snow sweeping down the mountainside for as much as 2–3km (1–2 miles), and destroying everything in its path. Large powder avalanches generate wind-speeds of more than 300kph (200mph), which can flatten trees even after the snow itself has stopped moving. Similarly, slab avalanches can range from limited slides of compacted snow to major slides, which strip whole mountainsides of any trees or buildings in their path. The really vulnerable slopes and gullies will have a distinctive lack of trees.

Assessing the risk

The risk of avalanches depends on the angle of the slope on which the snow collects and on the depth and the condition of the snow that is present. Of these factors, only the angle of the slope can be measured with any degree of accuracy. The most vulnerable slopes for slab avalanches are those of between 35 and 40 degrees. Such avalanches rarely occur on slopes of less than 20 degrees and greater than 50 degrees because the former are too flat and the latter are usually too steep

Left: A dramatic loose-snow avalanche that took place on Mount McKinley in Alaska, USA. This type of avalanche can wreak havoc on alpine communities.

to accumulate large amounts of snow. This said, wet, melting snow can slide down slopes that are even shallower than 20 degrees and slabs of new snow, which often build up temporarily when snow falls in calm conditions, can become dislodged on slopes that are steeper than 50 degrees.

Avalanche prediction is becoming more and more complicated because the risk in the Alps is increasing with the shift from agriculture to tourism. The decline in upland grazing in the summer means that the cows do not crop the grass short and long grass is not only beaten down by the snow but can ferment slowly during the winter months, making the surface slippery. Both these factors make it more likely for snow to slide off when the spring arrives.

Warnings about the depth and state of snow are prepared and distributed around the world. In the Alps the risk is measured on a scale of one to five. Even at the lowest level there is never a zero risk of avalanches, because, although there may have been no new snow for many days or even weeks, and temperatures may have been reasonably steady, there could still be isolated places on steep slopes where the snow has become unstable. High-risk levels give a broad indication of the dangers but should not be used to make assumptions about any particular slope. Although the avalanche services in many countries draw on large numbers of trained observers to determine the level of risk, the official figures should only ever be used as a guide.

Surviving an avalanche

Recent Swiss data provide a complicated and pessimistic picture. Initial survival rates following avalanches were surprisingly high, with a survival rate of 94 per cent in skiers dug out within 15 minutes. Of the fatalities, only a quarter were asphyxiated, while the others died from injuries sustained in the avalanche. After that, the survival rate plummets to only 30 per cent after 35 minutes. It then remains virtually steady until around 90 minutes, and then falls from a figure of 27 per cent to only 3 per cent surviving more than two hours.

These new data contain a set of vital messages. First, survival is critically dependent on being dug out by someone nearby; the rescue services will not necessarily get there in time. So if you are part of a group crossing an unsafe slope, make sure you are all well spaced out so that, in the event of an avalanche, you will not all be buried but will be able to dig each other out of the snow.

Second, an air pocket around your face will enhance your survival chances. Although this is likely to be a matter of chance, covering your face with your hands and arms may be the best way of protecting yourself.

The havoc wreaked in centuries-old hamlets across the Alps during February 1999, when over 60 people died in their shattered houses, is a tragic reminder of the devastating consequences of avalanches. These catastrophic events show that, while off-piste skiers may run the greatest risk, when the big snows come everyone in the mountains is in danger. Walls of snow hundreds of metres wide and up to 5m (16ft) high crushed chalets and apartments like matchboxes and gave their occupants no chance of escape. The lucky ones were protected by walls in the centres of buildings, but many were entombed in snow, which was compressed like concrete under the force of the headlong rush down the mountain. Desperate rescuers laboured night and day to find survivors, but, with few miraculous exceptions, it was too late to do anything for those who had been hit by the full force of the winter of '99.

MOUNTAIN VEGETATION

MOUNTAIN VEGETATION VARIES GREATLY, FROM THE
LUSH RAINFORESTS OF THE TROPICS TO THE SPARSE
AND STUNTED ELFIN FORESTS IN TEMPERATE ZONES.

*Above: Flourishing vegetation in a
mountainous region of Chile.*

Mountain vegetation can be separated into three categories with increasing altitude – the montane, subalpine and alpine zones. Plant life in these categories varies considerably around the world but, as a general rule, vegetation in the montane zone is very similar to that of the northern forests because it has the same sort of cold snowy winters in which coniferous tree species thrive. At the uppermost limit of the montane zone is the tree-line – the point on the mountain at which the forests stop growing and are replaced by occasional scrubby, stunted trees. Above this line is the subalpine zone, which is usually a relatively short transition area where trees peter out altogether. Higher still is the alpine zone, which continues until the altitude is such that the ground is under permanent snow cover and no vegetation can survive. This is the most distinctive zone of many mountain ranges as it is here that unique alpine plants are found.

THE MONTANE AND SUBALPINE ZONES

At high latitudes the climate of the lower mountain zones equates to that of the tundra, with long cold winters and warm summers. As such the vegetation on the mountains is not very different to that of the environment as a whole – forests of thin evergreen coniferous trees. In contrast, the forests in the montane zones of temperate mountains are thicker and stronger because they receive more sunshine and warmth during the summer months.

Eventually, the upper limits of the forest are superseded by what is known as krummholz or elfin forest (a dense thicket of low, contorted, horizontally spreading trees and shrubs) or by an area of sparsely distributed, small, upright trees. Coniferous krummholz is particularly widespread in north temperate mountains and in Colorado and Wyoming. The trees become even more scrubby in the subalpine zones, greatly influenced by strong winds. These combine with desiccation to deform trees and produce distinctive growth patterns, known as flagging, where most branches grow on the sheltered leeward side of the trees.

In the tropical regions, cloud forests (evergreen forests that grow at a level where clouds regularly form) carpet the mountain slopes of Africa and South America, Indonesia and New Guinea. The trees are constantly dripping wet and are laden with mosses, orchids and other epiphytes. The upper levels of the forests are cooler and contain a mixture of temperate and tropical species, so oaks are often interspersed with tree ferns, palms and bamboo. The higher levels of the Andes are often arid and various species of Puya (a member of the pineapple family renowned for its large spiky flowering stems, wax-like blue and green flowers and bright stems) grow at levels as high as 4000m (13,100ft). In East Africa, above the montane rainforest, there are characteristic zones of bamboo and heath species.

WHAT IS THE ALPINE ZONE?

Conditions in the alpine regions are severe: the temperatures remain low and the intensity of the sunlight increases sharply, which means there are huge variations in day and night temperatures. In fact, at low latitudes, the alpine climate is often referred to as 'summer every day and winter every night'. The climate is also very varied with a great deal more snow and rainfall on windward slopes than on sheltered leeward slopes, and dramatic differences in temperature between sunny and shaded areas.

The alpine zone begins roughly where the 10°C (50°F) isotherm falls in the warmest month of the year (an isotherm is a line on a map linking places of equal temperature). The altitude at which this occurs increases with decreasing latitude: it ranges from a mere 2000m (6500ft) in the Alps and the Canadian Rockies at 49–54°N, to around 3900m (13,000ft) in Sikkim in the Himalayas at 20–30°N, and to a lofty 4500m (14,800ft) in the Andes on the equator.

SURVIVING IN THE ALPINE ZONE

Various species are able to exploit the niches offered by the wide range of meteorological conditions found on mountains. There are general types of vegetation, such as the low-shrub communities usually found above the upper forests; mountain grasslands and heath; distinctive communities on rocks, cliffs, screes, moraines and exposed summits; and communities that exploit bogs, streamsides and areas where snow is melting.

At the highest reaches of the alpine zone, cold-tolerant mosses and lichens exploit sunny, sheltered spots, which have the highest daytime temperatures. These much sought-after locations are found where the snow is melting quickly enough to permit growth and slowly enough not to drown young plants. Capable of withstanding desiccation in winter or summer, some lichens live for over 1000 years. In an environment where longevity is the key to success, these species are among the great survivors.

A REFUGE FOR UNIQUE SPECIES

Many mountains around the world are isolated from other similar alpine areas. As a result, plants have evolved in isolation without interbreeding between related species or forms, and they have their own distinctive characteristics. These endemic species, which are found only in a single area, often occupy specific niches in the environment where they avoid competition. This is particularly true of various regions in Europe, where isolated mountains became refuges for certain species during the ice ages. Mount Olympus in Greece, for example, has as many as 20 endemic species, which only occur on its slopes.

In spite of the isolated nature of many alpine areas, there are close genetic relationships between many of the species found in these different places. This suggests that the present alpine floras of the temperate zone are the remnants of circumpolar cold-tolerant plants that used to be much more widespread during colder periods in the Earth's past climatic history.

Left: Nevado de Huascarán in the Cordillera Blanca of the Peruvian Andes. In the foreground is the verdant vegetation of the montane zone rising gradually to the sparser subalpine and alpine zones.

LIVING IN THE MOUNTAINS

EXTREME WEATHER, COMBINED WITH LESS OXYGEN IN
THE AIR AND HARMFUL ULTRAVIOLET RAYS, MAKES
LIVING IN THE MOUNTAINS A REAL CHALLENGE.

People who have lived in the mountains all their lives are very well able to cope with adverse mountain weather conditions. However, when lowlanders venture up the slopes they often suffer from altitude sickness and, although they may become more accustomed to the environment, they never develop the native highlanders' ability to cope with conditions.

MOUNTAIN PEOPLE

We all react in different ways to the effects of great heights. People born at high altitudes and who live there all their lives have much higher capacity for oxygen absorption than lowlanders. Interestingly, however, mountain people do not retain this capacity when they descend to low altitudes. Mountain dwellers are sometimes portrayed as being barrel-chested and short in stature. Indeed, some people who live in the Andes do have unusually large chests and lungs, which enable them to extract more oxygen from the air. Yet this is not a universal feature of such populations – the Sherpas of Nepal, for example, do not have large lungs and yet they are still capable of extracting large amounts of oxygen from the air.

ALTITUDE SICKNESS

Anyone who moves quickly to an altitude of 3000m (10,000ft) or higher, whether going on a skiing holiday in the Rockies, trekking in the mountains of Nepal or making a business trip to La Paz in Bolivia needs to take care when they first arrive at their destination. This is particularly important if you are doing strenuous exercise because if you do too much immediately, you may start to feel the effects of reduced levels of oxygen in the bloodstream. This condition is known as hypoxia and symptoms can include an increase in heart rate, shortness of breath, headaches and possibly faintness.

In some cases, the effects of altitude can be more severe. Acute mountain sickness (AMS) can develop in just a few days and symptoms include loss of appetite, headaches, tiredness, nausea, vomiting, dizziness and sleep-disturbance. Without treatment, AMS can become life-threatening, either because of the build-up of excess fluid in the lungs (pulmonary oedema), which causes breathlessness, or due to fluid accumulating in the brain (cerebral oedema), which causes confusion, drowsiness and in the worst cases can make the patient lapse into a coma. The prevalence of AMS rises from approximately 10 per cent at an altitude of 3000m (10,000ft) to more than 50 per cent at heights of 4000m (13,100ft), with some 1–2 per cent of people experiencing more serious complications at this higher level.

Right: A yak herder's daughter collects firewood and watches the herd on the Nimaling Plateau, Ladakh, India.

At an altitude of 5000m (16,400ft) the air pressure is only half of what it is at sea level and so there is only half as much oxygen available. Furthermore, our blood is less efficient at carrying oxygen at high altitudes. The combination of these effects makes us breathe much more rapidly. This increases the amount of oxygen in the blood but reduces carbon dioxide levels, which can be responsible for causing AMS. So it is best to take it easy when first staying somewhere at high altitude until you become acclimatized. Drink plenty of water and try to avoid alcohol because dehydration caused by the dryness of the air can make the symptoms of hypoxia much worse. If the symptoms persist, the best cure is to descend to lower altitudes. Moreover, wherever possible, sleeping at levels of less than 2500m (8200ft) helps greatly, hence the mountaineers' proverb 'climb high, sleep low'.

CAN LOWLANDERS ACCLIMATIZE?

Adults who move to high altitude and live there for many years never develop the capacity to extract large amounts of oxygen from the air. Children, on the other hand, who move when they are young, can eventually extract as much oxygen as native mountain dwellers.

And so the assumption that people totally acclimatize to altitude after a few days is unfounded. Initially, the response is good and after a few days, many people adapt because they begin to regulate their breathing more effectively. After this, however, there is little improvement and tests show that people's ability to work remains well below sea-level values. This capacity does not improve even over many months. However, on return to low altitudes it returns to normal immediately.

WHY ARE ULTRAVIOLET RAYS HARMFUL?

Mountain regions often get much more sunshine than the lowlands around them. This can have serious implications for recreational activities, especially skiing.

Sunlight contains two different forms of ultraviolet radiation, UVA and UVB. UVA, which makes up 98 per cent of all ultraviolet radiation, does relatively little damage to the skin (although many suntan lotions protect against it anyway). UVB, which makes up the remaining 2 per cent, is very dangerous because our skin is between 100 and 1000 times more sensitive to it than to UVA. It causes skin inflammation and can damage our genetic material (DNA). UVB is only partially absorbed by the ozone in the stratosphere, which is why it is important for us to use protective suntan lotion whenever our skin is exposed to the Sun. This need for protection is less extreme at low altitudes because tiny quantities of ozone that are present in the lower atmosphere beneath the mountain tops, together with dust and aerosols, continue to absorb any of the damaging UVB rays that get through the ozone layer.

PROTECTING YOURSELF FROM THE SUN

As a general rule of thumb, the UVB 'dose' people receive as they go higher in the mountains increases by about 4 per cent in terms of the sunburning power with every 300m (1000ft) in altitude. At 3000m (10,000ft) the UVB dose is 50 per cent more than that received at sea level. In addition,

the high reflectivity of snow means that at this altitude you may receive two to three times the dose experienced when sitting on a beach.

People who have lived at high altitudes for many generations tend to have skin types that tan easily and provide some protection from the damaging effects of ultraviolet rays. However, if holidaying in mountain regions, it is best to use plenty of sunblock, and also to wear a hat and protective clothing, especially if you have fair skin that has the tendency to burn easily in the sun. The dryness of the air and increased wind speed make skin protection on such occasions of paramount importance.

Above: Mountaineers, who have not lived at high altitudes their entire lives may be vulnerable to altitude sickness and must be careful when they first scale new heights.

THE THREAT OF TOURISM

MILLIONS OF US GO SKIING, CLIMBING AND WALKING
EVERY YEAR, BUT HOW LONG CAN WE EXPLOIT THE
MOUNTAINS BEFORE WE LEAVE AN INDELIBLE MARK?

Recreational activities in mountain areas are booming, but the vulnerability of alpine ecosystems and their slow regeneration mean that the current rapid pace of development may be doing irreparable harm.

THE NATURE OF THE THREAT

As more and more people are going to mountains around the world to take advantage of their unique climate, the pressures on the local environment are becoming unsustainable. The problems range from oxygen cylinders and garbage left by mountaineers on Mount Everest, to excessive forest clearance to provide cooking fuel for trekkers in Nepal and burgeoning ski resorts spreading across the Rockies and the Alps. In the latter, the number of visitors has risen fivefold in the last two decades and, as a result, Europe provides a model for global challenges, which will have to be faced in the future.

The demands of skiing highlight the challenges of exploiting a popular activity without exhausting natural resources. The development of new resorts, or the expansion of existing ones, often involves redesigning slopes high in the alpine zone where unique mountain vegetation thrives, cutting down trees at lower levels to provide ski runs to lower resorts, building roads for increased vehicle access and covering the mountainsides with a network of ski lifts. There are now over 40,000 ski lifts in the Alps, capable of transporting 1.5 million people an hour up the slopes, and the number is rising fast. In addition, there are over 5000 snow-making cannons, and there is a rush to install many more to counter the threat that global warming will make the snow cover less reliable (see box, right). This equipment smothers meadows at lower levels in an unnaturally long-lasting deck of snow, which often delays the growth of grass and prevents spring flowers from coming out on time.

These wintertime pressures are compounded by a desire to maximize the return on building investment by promoting activities in summer as well. While the traditional activities of walking and climbing exert some pressure on the environment, new pursuits such as biking, paragliding and four-wheel-drive 'off-road' racing are generating a much heavier burden.

The threat to the Alps is heightened because the mountains are at the hub of Europe. Air pollution, caused by industry in nearby lowland areas and by heavy traffic on the 400,000km (250,000 miles) of roads running through the Alps, is producing a decline in air quality that is damaging the forests. In the long term, global warming poses an additional threat to sensitive vegetation, which will migrate up the mountains as temperatures rise. In effect, these species will face extinction when they run out

Right: The wreckage of a campsite situated near Biescas in northern Spain, where 83 people died when a flashflood roared down from the Pyrenees on 7 August 1996.

86 What is the alpine zone?

Above: A network of chairlifts and cabines carrying skiers up the mountains near Les Menuires in the Trois Vallées ski region of the Haute Savoie, France.

FUTURE OF WINTER SPORTS

Most exploitation of the mountains revolves around winter sports, so the scale and pace of development depend on how reliable future snowfall is perceived to be. With the prospect of global warming, one might expect snowfall levels to be dropping. In practice, the figures for high-altitude resorts in both the Alps and western North America show little evidence of any decline (see below). What appears to be happening is that, during the winter, any rise in temperature has been matched by a rise in snowfall. During the rest of the year, warmer drier summers have been the principal cause of the recession of the glaciers. This means that, while lower resorts (typically the old traditional alpine villages) may face an uncertain future, the newer higher developments do not have to worry too much about climatic change at the moment. The danger is that the pressure to invest in resorts that appear to have a better long-term future will only compound the environmental problems already pressing on the highest and most fragile of mountainous regions.

of mountain to retreat to at high levels. In the meantime, any increase in the diversity of species at higher levels will simply be a result of this migration not a sign of regeneration.

INCREASED FLOODING

The damage to the alpine uplands and forests affects not only the local environment, but has major implications for the safety of those living in the lowlands. The chances of damaging flash floods are increasing because of the accelerated run-off of rainwater from the upper slopes stripped bare of vegetation, and the impaired ability of the forest to absorb this run-off. This is a hazard for those who

enjoy camping in the mountains in summer – riverside campsites have been the scenes of many disasters in recent years in places as far apart as the French Alps, the Pyrenees, Korea and the Rocky Mountains.

REPAIRING THE DAMAGE

Reversing this trend will not be easy. The slow growth of fragile mountain vegetation in the alpine zone and the demands of tourism will limit progress. Any plans to re-establish ground-cover require not only financial investment but also improved knowledge as to what different types of vegetation need to survive in tough conditions. Studies of alpine forestry have shown that, even at low levels, different trees prefer very different habitats. These demands will be even harder to meet if the popularity of winter sports declines and the economic base for funding the repairs is undermined.

SNOW DEPTH AND COVER IN THE ARLBERG

Records kept in Austrian mountain villages for the last 100 years, of daily snow depth and the number of days with snow cover, provide a good measure of the amount of snow that fell each year. In the Arlberg region, at an altitude of around 1600m (5300ft), there is little evidence of any significant trend during the 20th century.

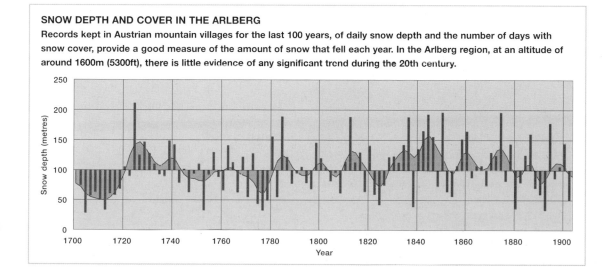

Global warming 184–5

MOUNTAIN GLACIERS

GLACIERS ARE VAST EXPANSES OF SLOWLY MOVING ICE THAT ORIGINATED FROM THE SUSTAINED BUILD-UP OF SNOW OVER MANY YEARS.

These grand and impressive features of the mountain landscape are an integral part of the scenery of upland areas, as well as being one of the most obvious indicators of the Earth's changing climate. They come in many different shapes and sizes: valley glaciers are huge frozen rivers of ice that form between mountain tops, smaller cirque glaciers are found in mountain basins and hollows, and the less-common piedmont glaciers build up at the bottom of mountain ranges.

Only about 1 per cent of the glacier ice on the Earth's surface is actually found within mountain ranges – the rest is locked up in the vast expanses of the ice sheets of Antarctica and Greenland. Nevertheless, glacier ice is an important part of the climate of the mountain regions and provides many insights into the current and past climates of these places.

HOW DO GLACIERS FORM?

Glaciers form when snow accumulates at high altitudes and recede when ice melts at low levels. Over time, the snow turns into ice in two ways. Firstly, it is compressed by subsequent layers of snow and eventually freezes, and secondly some of it melts in the summer and sinks down to lower levels where it refreezes as ice. The overall size and dynamics of a glacier are defined in terms of the mass balance and can be determined by looking at its lower reaches, which are known as the snout. A growing glacier has a positive mass balance, where more snow is accumulating at the top than melting at the bottom, with the result that the snout starts to move down the mountain. A receding glacier has a negative mass balance because the losses at low levels are greater than the ice that is being formed at high levels.

HOW FAST DO GLACIERS MOVE?

The rate at which glaciers flow downhill depends on two factors: precipitation rates at high levels and the gradient of the mountain. In the Southern Alps of New Zealand on the steep western flank of the Mount Cook range, precipitation over the Fox and Franz Josef glaciers measures some 10–15m (33–49ft) per year and much of this falls as snow at high altitude. The average velocity of the glaciers is 700m (2300ft) a year, although heavy rain acts as a lubricant, which means that they can move as much as 2500m (8200ft) per year.

The ice inside a glacier flows at different rates. The central upper part flows more quickly than the edges, which are impeded by the rocks they pass over, and the bottom, which drags along the bedrock. This causes huge stresses and strains within the glacier, which cause the ice to fracture near the surface and form deep fissures (crevasses). In the winter these are often covered by layers of snow and so become potential death traps for the unwary. If you wish to ski or trek in the magnificent splendour of glacier country make sure you always take a guide who knows the terrain and will be able to warn you about any possible risks that lie ahead.

These two pictures of the Horn and Waxeggkees glaciers in the Austrian Tyrol (top: 1921; above: 1994) provide clear evidence of the dramatic retreat in European glaciers, in which about 30 per cent of the area and 50 per cent of the volume has been lost since the middle of the 19th century.

Above: This is an impressive aerial view of an enormous glacier that is situated in the Chugach mountain range in Alaska, USA.

THE HISTORY OF GLACIERS

As a glacier slides down the mountainside, it leaves a trail of debris behind it, which is known as a moraine. Made up of rocks, mud and ice, which were mixed together by the constant grinding glacial movement, a moraine can reveal much about the glacier's history. Studies of moraines in the European Alps, Iceland and Scandinavia suggest that there were marked glacial advances in the 1590s, 1690s and 1810s – decades that form part of the Little Ice Age. Indeed this evidence is one of the strongest arguments for the reality of the Little Ice Age in Europe as cool wet summers are more significant than cold winters in glacier growth. Elsewhere, detailed evidence of glacier fluctuations prior to 1800 is fairly thin on the ground.

On longer timescales there is some evidence that glaciers around the world have expanded and contracted approximately every 1000–2000 years. The most notable early global advances occurred around 2300BC, 1200BC and AD800. All these periods coincide with times when many of the early civilizations were in recession and so can be interpreted as evidence of the impact of climate change on human history.

RECENT GLACIAL RECESSION

The recession of glaciers during the 20th century has been a global phenomenon. It has, however, varied from place to place and over time. The response of individual glaciers has depended on whether that place has experienced more or less warming than the rest of the world, and whether precipitation has risen or fallen. For instance, the prolonged period of retreat was interrupted by the cool wet summers of the 1910s and 1920s, and again in the 1970s and early 1980s in the Swiss Alps. In recent years, some glaciers in Norway and the Southern Alps of New Zealand have been advancing, despite the fact that temperatures have been rising, because snowfall amounts have more than compensated for this warming trend. Elsewhere glacier

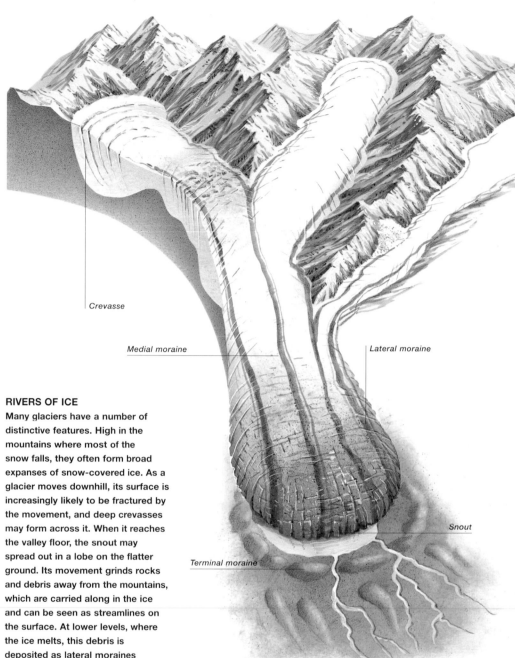

Crevasse

Medial moraine

Lateral moraine

Terminal moraine

Snout

RIVERS OF ICE

Many glaciers have a number of distinctive features. High in the mountains where most of the snow falls, they often form broad expanses of snow-covered ice. As a glacier moves downhill, its surface is increasingly likely to be fractured by the movement, and deep crevasses may form across it. When it reaches the valley floor, the snout may spread out in a lobe on the flatter ground. Its movement grinds rocks and debris away from the mountains, which are carried along in the ice and can be seen as streamlines on the surface. At lower levels, where the ice melts, this debris is deposited as lateral moraines alongside the glacier and as a terminal moraine beyond its snout.

levels have remained relatively stable in southern Russia/Kyrghizstan and have receded greatly in the European Alps, the Pacific coastal ranges of North America and parts of the Himalayas.

WHAT CAN LICHEN TELL US?

As glaciers move, they scour the rocks below clean of any living matter. When they start to recede, the pristine surfaces are gradually recolonized. Lichens are among the first colonizers growing at the slow rate of several centimetres per century. By working out how long they have been there, scientists are able to determine for how long a period the surface has been exposed to the elements. They can calculate this for the last 200 years to an accuracy of plus or minus five years.

The Little Ice Age 120–1

Global warming 184–5

MEDITERRANEAN REGIONS

CLIMATE OF THE MEDITERRANEAN

IN SUMMER, MILLIONS FLOOD TO THE MEDITERRANEAN
AND AREAS WITH A SIMILAR CLIMATE TO ENJOY THE
LONG WARM DAYS AND MILD BALMY NIGHTS.

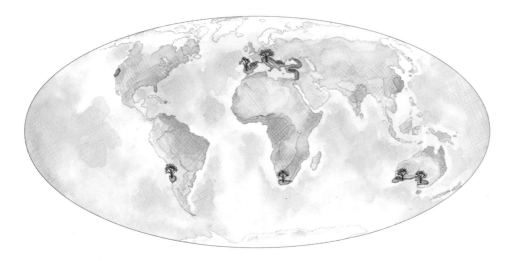

PALERMO, ITALY

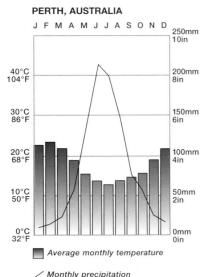

PERTH, AUSTRALIA

■ Average monthly temperature

╱ Monthly precipitation

Sandwiched between the temperate climes and the tropical regions, areas with a mediterranean climate enjoy mild winters and hot summers. This pleasant weather bestows benefits on their inhabitants and makes them sought-after destinations for tourists.

Above: This dry farmland in the Algarve region of Portugal, parched by summer heat, is typical of the poor soil conditions in many Mediterranean areas.

DOTTED AROUND THE WORLD

The term 'mediterranean climate' is reserved for a narrow band between other more extensive climatic zones and as such is confined to a few limited areas around the world. These consist of California, central Chile, the Cape of South Africa, parts of southern Western Australia and South Australia extending into Victoria, and the countries around the Mediterranean and Black Seas, stretching some 3000km (1900 miles) inland into Europe and Asia.

The mediterranean climate is transitional in a special way because it is controlled by mid-latitude westerlies in winter and by subtropical anticyclones during the summer. This means that the mediterranean winter is equivalent to a mild temperate winter, with average January temperatures falling to 10–15°C (50–59°F). In the summer they rise to around 21–27°C (70–81°F), with the daytime highs similar to subtropical regions at around 28–38°C (82–100°F). Temperatures can sometimes reach the high levels of dry tropical areas, for example when a plume of hot air extends northwards from the Sahara across the Mediterranean. This can happen anywhere from Antalya in Turkey to Seville in Spain, where temperatures can rise to 45°C (113°F), which is too hot even for the most dedicated sun-worshippers. In other parts of the world, hot desiccating winds can blow in from desert areas – for example, the Santa Ana in California, or the Berg in South Africa – bringing a sharp rise in temperature and a far greater risk of forest fires.

COASTAL INFLUENCES

The close proximity of the oceans in most mediterranean-type areas means that they exert a strong influence on temperature patterns. There is a rapid rise in daytime highs as you go inland from the Pacific in Chile, where between Valparaiso and Santiago the average daytime high in January rises from 21°C (70°F) to 30°C (86°F); and in California, where there is an even more extreme contrast in going from San Francisco to Fresno in July, rising from 22°C (72°F) to 38°C (100°F). Mediterranean-climate locations are therefore not only sandwiched between temperate areas and the tropics, but also sandwiched between the ocean and the mountains. The climatic effects are compounded by the presence of cold upwelling currents just offshore, which mean that when the wind blows onshore, the coastline is plagued with dense sea fog.

SUMMER AND WINTER CONTRASTS

The transition between summer and winter usually comes quite suddenly. In the Mediterranean the switch occurs in late October as the high-pressure system above the Azores declines and its influence over the Mediterranean ceases. By this time the Eurasian land mass to the north is cooling rapidly while water temperatures across the Mediterranean basin remain high, sparking off a sudden upsurge in storminess. During the winter, the Mediterranean spawns a large number of depressions – a process that is known as cyclogenesis. The most frequent sources of these storms are the Gulf of Genoa and the southern Ionian Sea

− the further east you go, the stormier it becomes. The storminess of the Levant peaks in mid-December and late January so, if you are visiting the Holy Land in winter, the best time to go weather-wise is around New Year.

Elsewhere the same sharp transition can be seen when the winter rains arrive quite suddenly in mid-autumn. By comparison, the waning of the winter season tends to be a more drawn-out affair as the storm tracks slowly move away to higher latitudes. So although there will be plenty of fine weather during the spring, it is often not until the end of May that the weather settles down and becomes really hot and dry. This puts into context the historical observation that, in ancient times, Mediterranean seafaring was restricted to the period of May to October.

IT NEVER RAINS BUT IT POURS

In an early 1970s pop song Albert Hammond bemoans the fact that 'it never rains in southern California: man, it pours.' Just how much winter rain falls in the Mediterranean, and in other places with a similar climate, surprises many people who live in rainy temperate climates. For example, Gibraltar gets more than twice as much rain as London from November to March. But it is not a case of there being many more rain days in the mediterranean but, more simply, when it does rain it does it in style ('man, it pours').

The same pattern applies elsewhere in the world. However, rainfall figures are lower in southern California than in the Mediterranean, and only from San Francisco

northwards is the rainfall comparable. If you are looking for lower rainfall in the Mediterranean, it is not a matter of going south, but of finding a rain shadow in the shelter of a large mountain. The best place is in the lee of Sierra Nevada in southern Spain, which also receives more sunshine – the Costa del Sol deserves its reputation of having a benign climate in winter. In regions of the eastern Pacific such as California and Chile, there are large annual variations in rainfall figures. These fluctuations are tied up with changes in sea-surface temperature patterns in the equatorial Pacific and, when there is a warm El Niño event, the rainfall is usually well above average.

Above: The view from the Greek amphitheatre in Taormina, Sicily, southern Italy.

Left: A typical Mediterranean scene: olive trees growing in Provence, southern France.

MEDITERRANEAN WINDS

FROM THE SUSTAINED MISERY OF THE MISTRAL TO THE SULTRY HEAT OF THE SIROCCO, THE WINDS OF THE MEDITERRANEAN MAKE THE REGION UNIQUE.

LOCAL WINDS
Surrounded by land masses, the Mediterranean is visited by a variety of distinctive winds, which exert a powerful influence on the climate of the region. In winter the majority of important winds are cold, and come from the northern half of the compass. In summer the hottest winds come from Africa or from the Middle East by way of the Black Sea.

WINTER WINDS
(A) Vendavales
(B) Tramontana
(C) Mistral
(D) Bora
(E) Gregale

SPRING AND SUMMER WINDS
(F) Etesians
(G) Khamsin
(H) Ghibli
(I) Sirocco
(J) Leveche

Local winds are an essential feature of the weather of the Mediterranean. They play a major role in the lives of the inhabitants and for the millions who holiday in the region throughout the year they can make all the difference between beautiful and miserable conditions.

THE STORMY MEDITERRANEAN

The winds in the Mediterranean are particularly extreme because the region is surrounded by land masses – Europe and Asia to the north and Africa to the south – which define the wind temperatures over land. As such, when the winds blow from different directions at different times of year, they can bring dramatically variable weather to the area. The frequent winter storms in the Mediterranean make the area wet and windy and pull air down from far-away places – for example, from depressions over North Africa. Where these winds come from and how they are funnelled into the Mediterranean basin makes a considerable difference to what the local climate is like.

In the summer, much of the region is covered by a large high-pressure area that extends up from the tropical North Atlantic. In the eastern part of the basin the high surface temperatures across Turkey and the Middle East to north-western India create a low-pressure region, known as a heat low, which is responsible for the hot dry conditions. However, the low-pressure systems can produce cooling northerly breezes for the islands in the eastern Mediterranean and so should be taken into account when choosing a holiday destination.

WINTER WINDS

The Mistral wind provides a very good example of how large-scale weather patterns can produce intense local winds. It is caused by extensive high pressure over the eastern North Atlantic and low pressure over Europe, which funnel air down from the North into the Rhône Valley and out into the Gulf of Lions in the south of France. During the winter months, this wind feels even colder because snow-covered upland areas produce a downward flow of icy air from the Alps and the Massif Central. This cold dense air sinks down into the valleys producing what is known as a katabatic wind. When this rushes down into the Rhône Valley, it can create wind speeds of up to 150kph (90mph) near the Rhône Delta.

Between the months of December and May there are about 26 days when the Mistral reaches speeds of 60kph (37mph) or more, and over one in three days when the wind speed exceeds 40kph (25mph). For the rest of the year there are far fewer intense episodes, but there are still 50 days between May and November when the wind speeds exceed 40kph (25mph). When these winds are blowing fiercely, they interfere with many aspects of life, making it miserable for all who live in the region. For instance, the Mistral makes the otherwise balmy winter climate of the French Riviera much less attractive. In summer, the dryness of the wind, combined with the lack of rainfall, greatly increases the risk of forest fires.

The Bora is a violent cold north-easterly wind that blows down from the mountains to the eastern shore of the northern Adriatic and in winter it produces violent squalls and gusts sometimes reaching 185kph (115mph). These conditions are most vigorous when there is a well-developed low-pressure system over the central Mediterranean and a cold front moving south-east over the Adriatic. This can pull cold air out of central Europe, to combine with the freezing air that drains down from the mountains of Dalmatia. Squally conditions often develop with great suddenness, which can bring unexpected snow to southern Italy (where the wind is known as the Gregale) and can even sweep across into North Africa.

Along the eastern coast of Spain a winter wind similar to the Mistral, known as the Tramontana, and other winds called the Llevantades and Vendavales bring stormy conditions during the winter. The Llevantades is responsible for gales, which sweep through Catalonia, producing particularly heavy seas in the region. The gales are caused by depressions that cross the Mediterranean between France and Algeria and are at their most dangerous in autumn and spring. The Vendavales produces strong south-westerly winds through the Straits of Gibraltar and up the eastern coast of Spain. These are associated with eastward-moving depressions from late autumn to early spring, which result in violent squalls and thunderstorms.

Above: Waves beating onto the shore of the island of Gozo, Malta in the southern Mediterranean.

Right: A vineyard at Godelleta, Valencia, Spain. Hot dessicating winds can make life difficult for farmers in the Mediterranean by damaging crops.

SUMMER BREEZES

The summer climate of the Mediterranean changes the character of the winds. Because low-pressure systems are not spawned in the region, the pattern of winds is defined by subtropical anticyclones and the heat low over Anatolia. So what matters is where the air is coming from and what happens to it on the way. For example, the hot southerly wind, known as the Sirocco in Italy, the Leveche in Spain, the Ghibli in Libya, and the Khamsin in Egypt (see far left), is most common in spring as the subtropical high-pressure region moves northwards. It draws hot, dry, dust-laden air from the Sahara, but picks up copious moisture on crossing the sea, which produces a disagreeable combination of heat and humidity. An additional hazard in spring and early autumn is that as it passes over the cooler waters of the northern Mediterranean it readily forms thick fog.

Perhaps the best-known summer winds are the Etesians, which blow from the north-west and north-east of the Aegean. These are caused by the heat low over Anatolia, and peak in August when the average wind speed is around 30kph (18mph) over the southern Aegean. Although they produce choppy seas, they moderate temperatures in coastal regions and on the Aegean islands, which is partly why these islands are such popular holiday destinations. Occasionally they may, however, cause violent thunderstorms, sudden wind and squalls. These are most likely to occur when the atmosphere is unstable as cold northerly winds at upper levels flow over hot surface air.

MEDITERRANEAN VEGETATION

HOT SUMMERS WITH LITTLE OR NO RAINFALL AND MILD AND WET WINTERS WITH CHILLY WINDS POSE VARIOUS CHALLENGES FOR LOCAL FLORA.

The flowering plants that provide the distinctive colour and fragrance of mediterranean regions have adopted various strategies to enable them to cope with the climate.

A HORTICULTURAL STOREHOUSE

Mediterranean plants are attractive to gardeners worldwide for many reasons (over and above their colourful appearance): their hardiness to extreme heat, cold, aridity and wetness and their ability to reduce evaporation and transpiration with their waxy or leathery leaves, which release aromatic oils. These oils make gardens fragrant and act as a basis for herbal recipes and remedies, which is why species such as lavender and rosemary have been common in gardens of northern Europe for at least four centuries.

There is a far wider variety of plants in areas with mediterranean-type climate in the southern hemisphere than in the Mediterranean itself. The land masses of the northern hemisphere have been in close contact throughout much of geological history and so many of the plant species in the Mediterranean have close cousins in North America. In contrast, the species in Australia, South Africa and South America have evolved separately for millions of years, each following its own course.

BULBS, CORMS AND TUBERS

One of the best ways for a plant to survive a long hot summer is by allowing all the above-ground vegetation to die back and to go into a state of dormancy, as seen in a bulb, corm or tuber. Then, revitalized by autumn and winter rains, flowering takes place in the emerging spring warmth. Many plants cope with the climate in this way, including crocuses, irises, lilies and tulips. They combine to make the countryside a riot of colour in spring.

MEDITERRANEAN HARDINESS

Human activities have left an indelible imprint on the vegetation of the Mediterranean and provided opportunities for hardy, drought-tolerant shrubs to become established. The plants that have culinary or horticultural attractions have been exported to colder regions for centuries. Their hardiness combats the blasts of cold northerly winds and occasional frosts and they can readily tolerate temperatures down to −10°C (14°F); some plants including the bay tree and Jerusalem sage can survive down to a frigid −15°C (5°F). In contrast, lowland areas in Australia rarely experience frost and so here the hardier specimens have only evolved in upland areas where frosts are more common.

CALIFORNIA CHAPARRAL

The evergreen shrub vegetation of California, parts of Arizona and northern Mexico is similar to the maquis of the Mediterranean, and both contain closely related species such as the Judas tree in Europe and the western redbud in the USA, or variants of the strawberry tree in Greece and California. Others, such as the many varieties of California lilac, which have become firm favourites in European gardens, have no Mediterranean counterpart.

The wetter uplands of California are dominated by forests. Around the Central Valley at low levels pine-oak woodland flourishes, and higher up in the Coastal Range blue oak woodland is replaced by mixed evergreen forest. The foothills of the Sierra Nevada support rich pine forests with western yellow pine, Douglas fir and, best known, although only limited in its extent, the giant sequoia, commonly known as the giant redwood. These forests provide some indication of how parts of the Mediterranean looked before they were deforested.

CENTRAL CHILE

The plants that grow in central Chile are similar to those that are found in California because the climate and geography of the two regions are alike. The striking shrubs of this region and of the temperate zone just to the south of it include the Chilean fire bush, the lantern tree and the potato tree, together with larger trees such as varieties of southern beech and the monkey puzzle tree.

Above: Lavender growing on the plateau of Valensole, Provence, France. Although a native of the Mediterranean, this attractive aromatic shrub has been widely grown in temperate regions for many centuries.

THE SOUTH AFRICAN FYNBOS

The fynbos is South Africa's version of the maquis. It is rich in brilliantly coloured flowers and shrubs. The fynbos is made up of 25 species of Protea – a genus of shrubs and small trees with colourful flowers, which can often be seen in florist shops. They have a remarkable defence strategy that enables them to survive drought conditions: their seeds remain dormant until triggered into germination by the smoke and heat given off by bush fires. Equally decorative are more than 50 different species of heather, which come in a much wider range of colours and forms than the heathers that are found in Europe.

Many of the bulbs and corms that are found in the fynbos have become firm favourites with gardeners around the world: for example, the green ixia, the harlequin flower and various forms of montbretia and freesia.

THE AUSTRALIAN EUCALYPTUS

Eucalyptus forests are the most distinctive feature of the Australian regions, which have a mediterranean climate. Eucalypts also extend widely into the more temperate and upland areas of Australia and Tasmania. Indeed the hardier species of these trees – for example, the snow gum, the spinning gum and the cider gum – have been introduced to many other parts of the world, where their fast growing characteristics and ability to withstand summer drought make them a valuable source of timber.

Above: A eucalyptus tree – a common feature of the Australian outback.

In Western Australia, jarrah covers over 1 million hectares (2.5 million acres) in areas where there is more than 1000mm (40in) of annual rainfall, while the karri tree forms forests in drier regions. These forests combine with a rich shrub flora including bottlebrushes, featherflowers, honeymyrtles and tea trees. Extending out to the arid heartland of the continent, the woods are dominated by eucalypts with an understorey of shrubs and herbs, eventually phasing into mulga scrub.

Left: Yuccas and Joshua trees found in the Joshua Tree National Park, California, USA.

THE MEDITERRANEAN DIET

THE MEDITERRANEAN DIET, RENOWNED AROUND THE WORLD FOR ITS CULINARY EXCELLENCE, HAS THE ADDED ADVANTAGE OF BEING EXTREMELY HEALTHY.

The Mediterranean diet is healthy because it is rich in fresh vegetables and fish, olive oil, cereals and wine. Production of all these components is intimately associated with, and depends on, the Mediterranean climate.

WHAT ARE THE BENEFITS?

The advantages of the Mediterranean diet show up clearly in the mortality statistics of countries such as Greece, Italy and Spain. An analysis of World Health Organization studies in 1990 showed that death rates from coronary heart disease in the Mediterranean were a half to one-third of those recorded in the USA and northern Europe. Even more striking is that the same low figures applied to the French, who, like their Anglo-Saxon cousins, have a diet that is high in animal fats. This is often termed as the 'French paradox' and has led to the assumption that the Mediterranean diet also contains some protective component that offsets the indulgences of fatty foods.

VEGETABLES

Since the time of ancient Egypt a wide variety of vegetables has been used as dietary staples in these regions. The Egyptians particularly prized onions and garlic because they kept them going throughout the day, were full of vitamins and mineral salts and were very easy to store for long periods of time. They also ate broad beans, cabbage, chicory, endive, leeks, lettuce, peas, radishes, shallots and water cress. Romans used the same range of vegetables, with particular emphasis on garlic, lettuce and onion. They may have been responsible for introducing these foods to Britain, where they were a staple part of the medieval diet, although garlic sadly fell from favour in later centuries.

OLIVE OIL

The olive tree is the symbol of the Mediterranean. The extent of its cultivation is a good measure of the climatic zone because its fruit-producing ability is limited by winter frosts in the north and by the need for autumn rains in the south. Otherwise it is peculiarly well suited to the Mediterranean climate because it is resistant to drought, immensely long-lived – with some trees living for over 1000 years – and a prolific producer of fruit.

Right: Klien Constantia Manor House and vineyard, Cape Province, South Africa.

22–3 The impact of volcanoes

96–7 Climate of the mediterranean

THE PRODUCTION OF WINE

When considering wine production, we need to extend our analysis beyond the boundary of what is normally regarded as the Mediterranean region and venture further north into France. This digression into temperate regions enables us to consider the weather sensitivity of some of the great wines of the world such as burgundy. Vines are hardy to winter frosts, which can be severe in this part of France. But, once vine buds break in April or early May these pampered plants are at the mercy of the weather: they are oddly sensitive to late frosts and so a mild winter leading to early budding followed by a late cold wave is the most difficult situation for wine-makers to handle. It then takes 100 days from the time of flowering in late May or early June for the grapes to reach full ripeness. Along the way every drop of rain, hour of sunshine and degree of additional warmth will affect the quality of the wine, but what precise mixture has the best results is impossible to say. As for disastrous weather conditions, hail at any time is a nightmare and so it is no wonder that wine-growers have been among the vanguard of those trying to modify the weather by seeding clouds to prevent hail.

By comparison, wine-makers in mediterranean climates have it relatively easy because of the limited geographical extent and less variable weather conditions, both within the season and from season to season. This sweeping assertion applies not only to Australia, California, Chile and South Africa, all of which produce high-quality wines, but also to southern Europe, Italy and Spain. The fascinating issue for wine drinkers is whether it is the ups and downs of the French climate that make French wines stand apart from the rest in particularly good years.

Above: An aged olive tree. These trees sometimes live for more than 1000 years and their fruit is mainly pressed to produce olive oil.

PAST WINE HARVESTS

The French historian Emmanuel le Roy Ladurie and colleagues have analyzed records of wine harvest dates in northern and central France, Switzerland, Alsace and the Rhineland spanning the period 1484–1879. This study provides an unrivalled insight into the vulnerability of wine production to the weather during the summer half of the year (April to September). The long-term variations must be viewed with caution because fashions for sweet wines have changed over the years, and with it the date of harvesting. However, fluctuations from year to year provide an accurate measure of the weather because they correlate closely with change in temperature. Wine harvests are also an indicator of the rainfall because the more rain there is, the cooler the temperature and the later the harvest. Just how much the weather can influence the production of wine in this region can be gauged by the huge range of harvest dates, which runs from 1 September in the blazing summer of 1556 to 24 October in the awful summer of 1816.

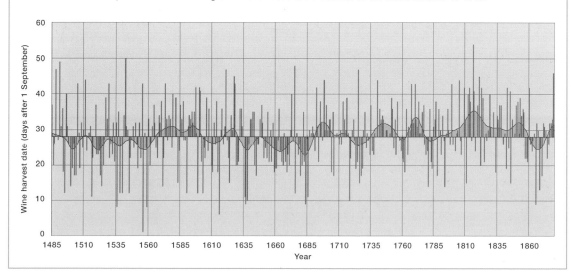

Cloud seeding 180–1

A HISTORY OF DEFORESTATION

THE BARREN LANDSCAPE OF THE MEDITERRANEAN IS A LEGACY OF EARLIER CIVILIZATIONS, WHICH DESTROYED THE FORESTS TO BUILD SHIPS AND MAKE WEAPONS.

Deforestation is not a new phenomenon but has been going on for centuries. The early alteration of the Mediterranean landscape and the way in which it has been dealt with since have important implications for the climate and environment of the region.

AGRICULTURE, COMMERCE AND WARFARE
Deforestation started during prehistoric times in the Mediterranean and the adjacent Middle East, together with parts of the Far East. It was a natural consequence of the development of agriculture in the more stable climate of the Holocene Period following the last Ice age some 11,000 years ago. The trees were cleared with stone axes, fires and domesticated livestock (notably sheep and goats), and this made it easier to use the land for agriculture, provided fuel and flushed out wild animals. Browsing goats have often been cited as a main cause of deforestation around the Mediterranean. It is, however, more likely that much of the mature forest at higher levels was used to provide the large timbers to build ships, for both the Greek and Roman fleets and to meet the needs of trade. Even by the time of Plato (429–347BC) the philosopher was expressing sadness at the disappearance of the lush forests. However, goats must share some of the blame because they prevented regeneration of the forest by continual browsing.

Below: The Battle of Lepanto, 7 October 1571. It is estimated that the building of the fleets involved in this last great battle between galleys required the felling of over a quarter of a million mature trees.

Once the trees had been cut down, heavy winter rains washed away the soil and this erosion reduced the chances of the forests regenerating, and ensured that only scrubby cover returned. The widespread adoption of the olive as a source of oil was another factor in the permanent alteration of the ecology of the region because the demand for olive trees grew and they became established. This tree is ideally suited to the climate of the region and has become an integral part of the landscape.

A SMALL PART OF LARGER CHANGES
Deforestation of the Mediterranean is just one small part of the changing face of the Earth over the centuries. Estimates suggest that, over the last 10,000 years, 5 per cent of the Earth's surface – 17 per cent of land surface – has been changed in some way because of human activity. This is mostly due to deforestation across the world and to salinization. These changes could be responsible for the reduction in global temperature that has taken place over the past 6000 years. This is because open ground is more reflective than vegetation and so the surface albedo has increased by 0.5 per cent – enough to reduce global temperature by 1°C (1.8°F).

CAUSES OF DROUGHT
Although the deforestation of the Mediterranean is only a small part of this overall change, it may have had important regional consequences. Over the years, drought has played a major role in social developments in the region. Rainfall declined around 4000 years ago, which ties in with the wider changes taking place at the time – mountain glaciers are thought to have expanded around 2500BC, but then receded to high levels around 2000BC. This coincided with the fall of the Akkadian civilization in ancient Babylonia. Another lengthy period of desiccation may explain the dark age of the eastern Mediterranean at the end of the 13th and beginning of the 12th century BC – this may have contributed to the collapse of the Hittite empire, Mycenae and Ugarit and to the enfeeblement of Egypt. Although other factors were often involved, drought undoubtedly made it more difficult for civilizations to progress because of the impoverished vegetation cover that it left. The upsurge of economic activity in the Middle Ages led to the last major clearance of forests.

MAQUIS AND GARRIGUE
The destruction of the former forest cover, with its stands of pine and evergreen oak, left the way open for the present day scrub (the taller denser variety is known as maquis and the thinner more widespread form, which survives on almost bare limestone, is known as garrigue). Centuries of cutting, firing and grazing have ensured that plant species able to generate quickly have become the dominant types. Many of the woody species have small thick leathery leaves, which reduce transpiration in the summer. They are aromatic and release oils, which may also reduce water loss and deter animals from grazing.

There are still some areas of the original forest remaining, although these are now intensively managed, and are a pale shadow of the earlier dense woodland. The more extensive regions of scrubby woodland are a mixture of evergreen spiny shrubs such as kermes oak, broom, heathers, juniper, rock rose and strawberry trees. Active growth and flowering take place in the autumn and often throughout the winter, and reach a peak in spring. In many places, where there is adequate soil to re-establish the ancient forest, the cover is now made up of olive groves.

MODERN ENVIRONMENTAL PRESSURES

The massive surge in tourism in the area has brought with it a new set of threats for the Mediterranean landscape. The population of the region has more than doubled in the last 50 years and the annual influx of visitors now exceeds 100 million, making the Mediterranean the most intensely exploited tourist area in the world. This means that many of the most vulnerable coastal areas are being overdeveloped, placing additional demands on limited water supplies and stripping the already barren landscape. As tourism grows, so too will the risk of soil erosion in the winter and of forest fires in the summer.

Left: Mud slides that struck the Campania region of Italy in May 1998 killed 101 people and destroyed the homes of at least 1500.

Left: A fresco depicting the month of December at the Castello del Buonconsiglio, Trent, Italy. Painted around 1400, it shows the traditional activity of chopping down trees for timber and firewood.

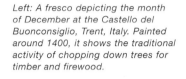
Sahel drought 162–3

TEMPERATE REGIONS

CLIMATE OF TEMPERATE REGIONS

MUCH OF THE EARTH'S POPULATION LIVES IN THE MOST MODERATE OF THE CLIMATIC ZONES – THE APTLY NAMED TEMPERATE REGIONS.

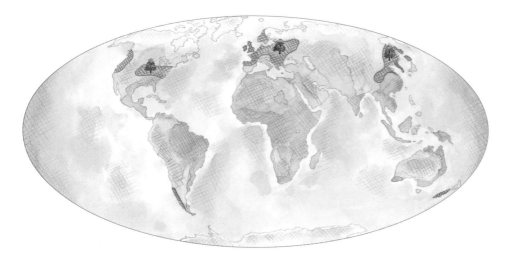

WESTERLY WINDS IN TEMPERATE AREAS

The weather in the mid-latitudes of both the north and south hemisphere is dominated by the prevailing westerly winds caused by the atmospheric circulation patterns. The path of these winds in the northern hemisphere is confused by the distribution of the continents around the world and so the temperate zone embraces a wide range of different climatic conditions. In the southern hemisphere, the situation is simpler.

The temperate zone covers many of the areas situated in the mid-latitudes, around 40–55°N and S. In the northern hemisphere it can be broadly divided into two categories: temperate oceanic, which is found on the western side of North America and Europe, and has cool wet winters and mild wet summers; and temperate continental, which lies on the eastern flank of North America and Asia, and has colder snowier winters and warmer wet summers. In the southern hemisphere, the absence of continents at these latitudes means that only temperate oceanic conditions are found here, and the only significant land areas affected are southern Chile and the South Island of New Zealand.

Below: A low-pressure system over the UK, bringing wind and rain. The line of clouds in the bottom right-hand corner is a cold front, and the swirl in the upper right an occluded front.

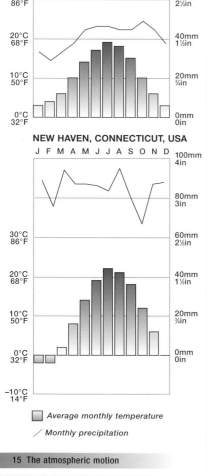

The mild wet winters and warm wet summers make the temperate zone ideal for agriculture and industry. The climate is stable and there are rarely ever extremes of temperature or of weather conditions.

Average monthly temperature

Monthly precipitation

WHAT DEFINES THE CLIMATE?

The climate of the temperate zone is governed by the transfer of energy around the globe. Warm air masses from the equator and cold air masses from the poles meet over the temperate zone and form mobile regions of low pressure, known as depressions, and less dynamic high-pressure systems, known as anticyclones.

Depressions and anticyclones are intimately linked with global ocean currents and the distribution of land masses. Depressions derive their energy from the oceans and their tracks are governed to a large extent by the distribution of land and sea. At the same time the westerly winds associated with the movement of depressions are a major factor in driving the ocean currents. Thus, there is a continuous feedback process, as depressions gain much of their energy from the warm ocean currents and in so doing generate the winds to drive the currents and replenish the supplies of warm surface water.

In terms of local weather, depressions cause storms, cloud cover and rainfall but keep temperatures up because they contain much energy. In contrast, high-pressure systems cause calm weather and clear skies, which push temperatures down in winter and up in the summer.

STORM TRACKS AROUND ANTARCTICA

In the southern hemisphere, the situation is simpler because of the near-absence of continents in mid-latitudes and the symmetry of Antarctica. The westerly winds are stronger and less varied from winter to summer, and the tracks of the depressions form a low-pressure region at 60–65°S. Their near-constant circular movement is mirrored in the circumpolar current in the southern oceans.

Above: A typical temperate scene – autumn at Hillside Farm, USA.

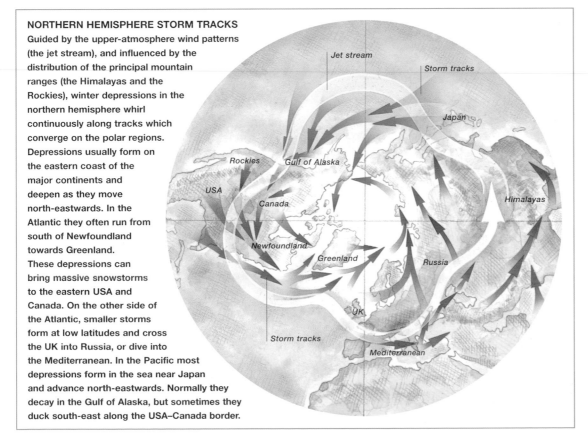

NORTHERN HEMISPHERE STORM TRACKS

Guided by the upper-atmosphere wind patterns (the jet stream), and influenced by the distribution of the principal mountain ranges (the Himalayas and the Rockies), winter depressions in the northern hemisphere whirl continuously along tracks which converge on the polar regions. Depressions usually form on the eastern coast of the major continents and deepen as they move north-eastwards. In the Atlantic they often run from south of Newfoundland towards Greenland. These depressions can bring massive snowstorms to the eastern USA and Canada. On the other side of the Atlantic, smaller storms form at low latitudes and cross the UK into Russia, or dive into the Mediterranean. In the Pacific most depressions form in the sea near Japan and advance north-eastwards. Normally they decay in the Gulf of Alaska, but sometimes they duck south-east along the USA–Canada border.

DEPRESSIONS AND ANTICYCLONES

SWIRLING DEPRESSIONS AND STATIC ANTICYCLONES,
CREATED BY DIFFERENCES IN PRESSURE, FUEL THE
UNSETTLED WEATHER OF THE MID-LATITUDES.

VILHELM BJERKNES (1862–1951)
After working for some years in Germany, Bjerknes became the leader of a group of Scandinavian meteorologists who worked in Bergen, Norway (the Bergen School) and in around 1920 he and his son Jacob developed the standard model of the mid-latitude depression. The essential feature of this model is that most weather activity is concentrated in narrow zones formed by the boundaries between warm and cold air masses. These are known as fronts – an analogy with the narrow battle fronts of World War I. The methods used to predict the movement of these fronts became central to weather forecasting and the basic Bergen model remains important today.

Depressions occur where warm air is rising up from the Earth's surface creating low-pressure systems. In contrast, anticyclones occur where cold air is sinking towards the Earth's surface creating regions of high pressure.

EARLY MODELS OF DEPRESSIONS
The first model of the extratropical depression was created in 1863, by Admiral Fitz-Roy, the first head of the Meteorological department of the British Board of Trade (the forerunner of what is now the Meteorological Office). He observed that depressions were normally composed of two air masses of different temperatures and moisture content. The warmer, more moist air mass originated in subtropical latitudes and the colder drier mass came from polar regions.

DEPRESSION: A MEETING OF FRONTS
One of the most important features of a depression is the sharp divide between hot and cold air masses, known as a front. Along this front, temperature, humidity, atmospheric pressure, and wind speed and direction change suddenly.

Sections of the front are named according to their motion. A warm front occurs where warm moist air rises and displaces cold air. This process is often likened to a rising conveyor belt and the zone of air is marked by clouds at increasing altitude up to some 1000km (620 miles) ahead of the surface front. A cold front is the reverse process with descending cold dry air pushing under warmer air. This usually occurs over a smaller distance and the convection process is more vigorous than that of the warm front. When the warm and cold fronts meet, a depression forms. This results in warm air rising, which creates extensive layers of cloud and precipitation – the characteristic weather features of a depression.

SECONDARY DEPRESSIONS
In its later stages the depression runs out of energy and begins to stagnate, but its trailing cold front can become the site for the next 'secondary' depression. In this way a family of several depressions can follow one another along similar tracks. This succession is often presented as an orderly process: rather like a troop of elephants trunk-to-tail in a circus ring, the depressions are shown girdling the globe. In real life it is a different story. Some secondary depressions are pale shadows of their forebears, while others rapidly grow into monsters that dominate the circulation of a significant part of the hemisphere.

BIRTH OF A DEPRESSION
The lifecycle of a depression starts where cold and warm air masses slide past one another. The warm air starts to rise forming a wave, which creates an area of low pressure known as a depression.

DEVELOPING DEPRESSION
The two fronts swirl around a central region of low pressure with a mass of rising warm air between them. Eventually, the cold front catches up with the warm front, using up the wedge of warm air between them to drive the depression through its development.

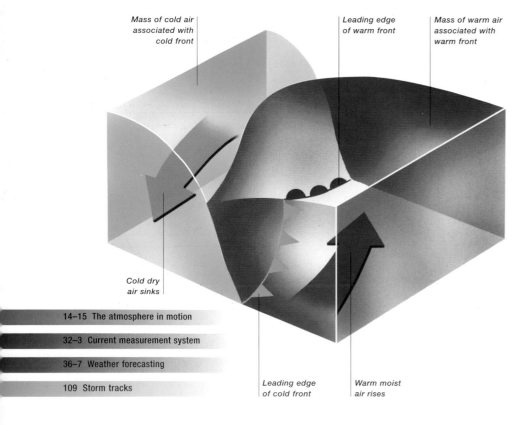

Mass of cold air associated with cold front

Leading edge of warm front

Mass of warm air associated with warm front

Cold dry air sinks

Leading edge of cold front

Warm moist air rises

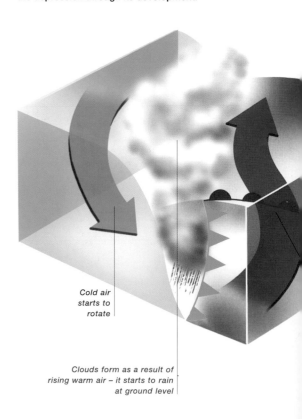

Cold air starts to rotate

Clouds form as a result of rising warm air – it starts to rain at ground level

PREDICTING HOW DEPRESSIONS BEHAVE

Although we understand the basic dynamics of a depression, we cannot explain or predict the behaviour of fronts. It is remarkable that the atmospheric transfer of heat around the globe should bring about such sharp transitions from hot to cold air (measuring no more than a few tens of kilometres in width) rather than a gradual spread of heat from equator to pole. In recent years fronts have been the subject of intense investigation combining aircraft, radar and surface measurements, satellite cloud observations and computer models. This work will help us to understand the unpredictable movement of fronts and why some fronts produce sudden depressions while others are quiescent.

ANTICYCLONES

In contrast to depressions and cyclones, anticyclones are regions of high pressure, which produce calm clear weather. They are more enigmatic than the organized depressions. They lack uniformity in their patterns of formation, growth and decay and are more irregular in shape and behaviour. Often they seem to be sluggish, passive systems, filling the spaces between their low-pressure counterparts, but they do affect some weather features.

Anticyclones can be divided into cold or polar continental highs, and warm or dynamic highs. Polar continental highs develop over northern hemisphere land masses in winter. They are created by intense cooling of the snow-covered surface, which gives rise to a shallow layer of dense very cold air. When the amount of cold air reaches a certain level

PRESSURE SYSTEMS
Warm moist air rising in depressions and cool dry air descending in anticyclones link surface weather systems with the jet stream. This connection between high and low levels means that the movement of depressions and anticyclones is controlled by troughs and ridges in the jet stream.

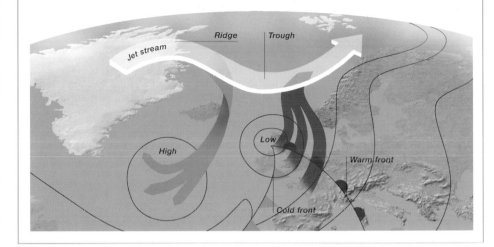

the air mass sweeps swiftly down to lower latitudes, bringing colder weather to large areas, especially in winter. In contrast, dynamic anticyclones are caused by large quantities of cold air sinking throughout the troposphere. These include the subtropical highs, caused by the sinking air of the Hadley atmospheric-circulation cell, and slow-moving mid-latitude highs, known as 'blocking anticyclones' because they block other weather systems causing spells of abnormal weather.

Below: A transmission tower collapsed during an ice storm that brought chaos to millions in eastern Canada and northern New England on 6 January 1998.

MATURE DEPRESSION
When the system reaches maturity, the cold front overtakes the warm front and is said to occlude. The resulting occluded front has no impact on the temperatures at ground level but warm air aloft may produce heavy precipitation.

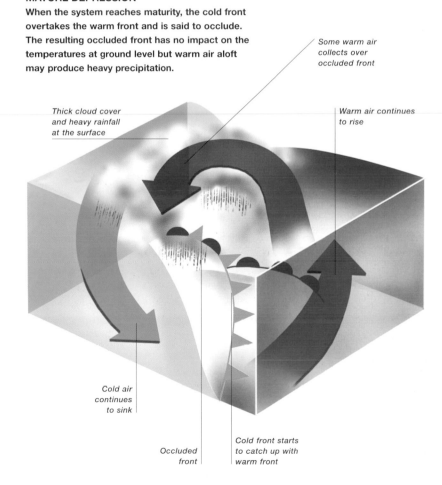

Some warm air collects over occluded front

Thick cloud cover and heavy rainfall at the surface

Warm air continues to rise

Cold air continues to sink

Warm air starts to rotate

Occluded front

Cold front starts to catch up with warm front

AGRICULTURE AND FARMING

FARMING OVER THE CENTURIES HAS ALWAYS BEEN AT
THE MERCY OF THE ELEMENTS, BUT THE CLIMATE OF
TEMPERATE REGIONS IS IDEAL FOR AGRICULTURE.

Temperate regions have always been particularly well-suited to farming and the long-recorded history of agriculture in the region reveals how farmers over the centuries have withstood the challenge of the weather. The temperate regions of Europe, north of the Alps extending from the Atlantic to the southern Urals, with year-round rainfall, provide ideal conditions for arable and mixed farming. The same applies to the north-east quadrant of the USA, parts of south-eastern Canada and the temperate areas of eastern Asia (for example, northern China, Manchuria, Korea and northern Japan). The difference between a good and bad harvest is not down to a single factor, such as the seasonal rains, but is usually due to a combination of factors over a long period of time.

SURVIVING THE WINTER

Although the winters can sometimes be extremely cold, especially in temperate continental regions, they have relatively little impact on the level of production because many crops are not planted until spring, and those that do have to survive the winter are dormant during the cold months. In fact, some plants benefit from the cold weather: vernalization is a process whereby trees, such as apple trees, do not come into bud until they have had a certain amount of cold weather. In other circumstances, cold weather slows down growth and prevents young shoots

from emerging until the threat of damaging frost has abated. Winter wheat is an example of a crop that benefits from scorching by extreme frost in the most severe winters. Indeed, farmers in the past often allowed sheep to graze on the emerging shoots in winter to ensure that they were not too advanced by early spring.

GOOD AND BAD YEARS

Successful cereal growing in north-western Europe and Canada appears to depend on a fine balance between adequate moisture and reasonable warmth, especially during harvest time. The combination of heat and drought can be as damaging as sustained cool, wet weather. In general, cool relatively wet summers produce the heaviest yields, as long as the harvest can be gathered in efficiently. The recipe for success tends to be plentiful rainfall and average temperatures until the end of June, and then reasonably dry and warm weather thereafter. Old European folklore reflects this with its definition of a good summer as being one that had wet weather until St Anthony's Day (24 June), followed by dry conditions. The adaptability of agriculture means, however, that only in the most extreme circumstances do yields fall dramatically.

THE WORST OF YEARS

One of the worst years for agriculture was 1879. The year began and ended with exceptionally cold winters with an unrelentingly cold wet growing season in between. In central England it was the coldest year in the past two and a half centuries. The temperatures every month were below average and the incessant rains produced the highest

Right: A hayfield in Estonia provides a traditional image of agriculture in temperate regions.

38 Wheat prices

108–9 Climate of temperate regions

summer soil-moisture figures since the late 17th century. The total output of British agriculture fell by around 20 per cent, while figures for the six principal crops in Ireland dropped by nearly 35 per cent.

The opposite extreme was the hot dry summer of 1976. Following a dry winter and spring, the summer was the hottest in central England in the last three centuries, as well as the driest. Agricultural production was hit across the board: cereal yields fell by about 30 per cent, while potatoes and root crops such as carrots and turnips were hit even harder. These declines in output are comparable with the losses that were suffered during the drought years of the 1930s in North America.

CEREAL PRICES

Where there are no accepted figures for yields or reliable weather records, the quality of the harvest can be judged by the fluctuations in grain prices. In bumper years prices were depressed, and when the harvest failed and yields were pitifully low, the prices rocketed. Various price series have been constructed by historians for places in north-west Europe, covering much of the period from the beginning of the 13th century. These provide useful insights into the fluctuations in the weather from year to year and highlight the most disastrous years. Furthermore, the cost of food was about 80 per cent of the cost of living for a medieval peasant and so the changes in the price of grain are an accurate barometer of social conditions. What these series show is that, while year-to-year fluctuations were largely due to the weather, long-term variations were the product of underlying economic factors.

RECENT CHANGES IN PRODUCTIVITY

Any study of recent variations in agricultural output as a result of weather fluctuations must take account of the dramatic rise in productivity since World War II. In Britain, for example, the output of wheat per hectare (2.5 acres) has risen from a little over 2000kg (2 tons) in the late 1940s (a figure that had remained almost constant since the late 19th century) to over 7000kg (7 tons) in the 1990s. Many other parts of the world have seen similar rises in productivity. Intensive use of fertilizers, herbicides and insecticides, combined with the development of improved varieties of crops and mechanization has contributed to this rise in production levels. These factors must be considered when assessing the correlation between productivity and weather.

Rising output may become more important if the world population continues to grow at the rate it is today. Any estimates of whether future food production can meet the needs of the world's growing population and accommodate changes in the climate must include assumptions about whether the rises in productivity that have occurred in recent decades can be sustained. Progress depends principally on developing improved varieties of grain, but also on whether food production will become more or less sensitive to fluctuations in the climate.

Above: Cumulus clouds above fields of straw stubble in Sussex, United Kingdom.

The Little Ice Age 120–1

Dust Bowl years 132–3

Future food production 176–7

THE IMPACT OF URBANIZATION

CITIES AND TOWNS THROUGHOUT THE WORLD ALL
HAVE AN INFLUENCE ON THEIR LOCAL CLIMATE – THE
BIGGER THE CITY, THE MORE DRAMATIC THE IMPACT.

Cities have a noticeable effect on the climate because their surface conditions are so very different to those in surrounding rural areas.

WHAT IS THE URBAN HEAT ISLAND?

There are a number of physical processes at work that make the climate of cities different from conditions in rural areas. Trees and vegetation are often replaced by buildings, which alter the surface roughness and hence windflow patterns. The brick and concrete canyons formed by the buildings and the asphalt of roads and parking lots absorb more of the Sun's heat than open ground. More heat is stored and released at night, which slows down the rate of cooling of the surface and holds temperatures up. Furthermore, vehicles and buildings use large amounts of energy, by burning fossil fuels and by consuming electricity respectively – these processes cause the temperature to rise and release large quantities of pollutants.

CHANGES IN THE WEATHER

Cities are substantially warmer than their surroundings, especially on calm clear nights when the difference can be more than 10°C (18°F) for large cities. On the whole, these temperature differences disappear when wind speeds rise above about 24–36kph (15–20mph), because they mix the air and prevent it from stagnating. Other weather conditions are influenced to a lesser extent: rainfall increases a little and wind speeds decrease noticeably. This said, high buildings can cause unpleasant swirling winds unless they have been designed with this aerodynamic problem in mind. The impervious surfaces of buildings also accelerate the run-off of rain and increase the risk of damaging floods, especially downstream of cities.

However, by far the most worrying aspect of city living is that air pollutants are built up during calm sunny conditions. High levels of pollutants are permanent and unwelcome features of urban areas; they include sulphur dioxide, oxides of nitrogen and particulates. A recent study carried out in Britain by the Committee on the Medical Effects of Air Pollution concluded that 24,000 people die each year in the country as a result of air pollution. Ozone, which is created when sunlight reacts with vehicle exhaust fumes to form photochemical smog, causes 12,500 premature deaths each year and sulphur dioxide claims another 3500 lives. Perhaps the most depressing statistic is that fine particulates cause over 8000 premature deaths. These are largely from diesel engines, which were until recently considered to be a more energy-efficient means of powering vehicles. Now the solution will have to involve better exhaust-cleaning technology.

IMPROVING LIFE IN THE CITIES

Some of the first, and most effective, environmental legislation was aimed at reducing the level of air pollution in cities. Following the 'Great Smog' in London in early December 1952, which caused over 4000 more deaths than usual, legislation was introduced in 1956 to control the emission of smoke from coal fires (the Clean Air Act). The effects of this legislation were dramatic – the amount of winter sunshine in central London rose by more than 70 per cent between the late 1950s and the 1980s. A similar

Right: Mexico City, at an altitude of 2300m (7550ft), with a population of over 20 million and abundant sunshine, suffers from some of the worst photochemical smogs in the world.

Left: Smog over Paris. In recent hot summers the combination of heat and air pollution has led the authorities to take urgent action to restrict the movement of vehicles in an effort to reduce smog levels.

story has occurred in the Los Angeles basin during the last two decades with the introduction of exhaust controls on automobiles, which has significantly reduced the incidence of photochemical smogs in the region.

These results show how we can address the threat of higher temperatures and worsening air pollution that are likely to occur as a result of global warming. Nearly half the population lives in urban areas and the proportion is rising. Many health problems are worsened by the effects of the urban heat island, especially in big cities in warm temperate regions (such as New York, Shanghai or Tokyo) and so solutions for these areas will need to be found soon.

COPING WITH HEATWAVES

When daytime highs rise above 35°C (95°F) the death rate in many mid-latitude cities rises sharply, especially if accompanied by high night-time temperatures. If average summer temperatures rise by 2–4°C (3.6–7.2°F), the death toll in cities such as Chicago or New York can rise by as much as several hundred. The widespread use of air-conditioning should keep these figures down, but there is greater concern for those developing cities, such as Cairo and Shanghai, without this luxury. When combined with other forms of local pollution, heatwaves make cities even more unhealthy places live. However, we will only really understand the impact of heatwaves if the incidence of exceptionally high temperatures increases.

The solutions lie in reducing vehicle emissions, promoting clean public transport and reducing power consumption. The pressures for local action reinforce the case for cutting down the release of greenhouse gases, which contribute to global warming. In addition, strategies designed to minimize the intensity of the urban heat island, such as planting more trees to create shade in suburban areas and

reducing the absorptivity of the surfaces of buildings (for example, by painting roofs white), parking lots and roads may be even more cost-effective.

URBAN HEAT ISLAND

The higher temperatures in major cities show up most clearly at night during calm clear conditions, but also affect daytime highs, as these figures for London show. Red indicates the warmest areas and purple the coldest. During the day and night, the centre of the city remains substantially warmer than the outskirts.

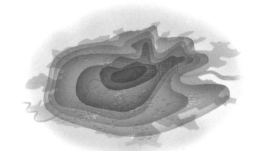

NIGHTTIME LOW TEMPERATURES IN LONDON

DAYTIME HIGH TEMPERATURES IN LONDON

RISK TO HUMAN HEALTH

The excess mortality during a severe heatwave shows up dramatically in these figures for a heatwave in New York in July 1966. This sharp rise is seen in the statistics for many large cities around the world, which experience heatwaves in summer.

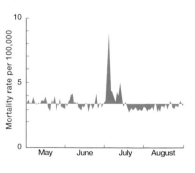

Climate and health 178–9

Global warming 184–5

UNUSUAL SPELLS OF WEATHER

A PARTICULARLY COLD SPELL IN WINTER CAN BRING
DISRUPTION TO THOSE LIVING IN TEMPERATE ZONES
– BUT WHAT CAUSES THESE UNUSUAL CONDITIONS?

The oceans and atmosphere transport energy around the globe and control weather patterns. As such, any variations in their circulation patterns will affect how much energy they transport to each region and may cause unusual spells of weather.

BLOCKING ANTICYCLONES

Circulation patterns in mid-latitudes are usually dominated by strong regular westerly winds. Abnormal spells of weather are caused by an unusual type of atmospheric circulation. The best-known example of this phenomenon is termed a 'blocking anticyclone'. This occurs when an area of high pressure becomes trapped between regions of low pressure and remains stationary for several weeks, causing a warm dry spell of weather.

To constitute a block, these cells of high pressure, sandwiched between normal westerly flow, must last at least 10 days, but they can last much longer. In the northern hemisphere they occur most frequently near the Greenwich meridian and in the eastern Pacific. In winter, they cause cold weather in north-west Europe and the eastern USA, and in summer they cause hot dry weather in both regions.

Blocking conditions occur when the upper-level westerly winds split into two well-defined branches, which extend over at least 45 degrees of longitude. Meteorologists are still not able to explain why the number of waves (named Rossby waves after the scientist who first identified them, see box, right) around the globe switches back and forth from three to up to six or why they can vary from only small ripples to exaggerated meanderings with isolated cells of high pressure, which cause blocking – the more meandering the waves, the more blocking that occurs).

In the northern hemisphere the distribution of land masses and of the major mountain ranges plays a major role in defining the path of the jet streams and hence the areas where unusual weather spells strike. (In the southern hemisphere blocking is less common because of the relative absence of land masses in the mid-latitudes.) The strength of atmospheric circulation in the northern hemisphere in any year seems to affect how the winds interact with land masses and mountain ranges. The stronger the winds, the greater the deviation around land masses and the more potential there is for blocking to take place. For example, when the wind speeds in the upper atmosphere are very high, strong blocking anticyclones are likely to build up downstream of both the Himalayas and the Rocky Mountains.

NORTH ATLANTIC OSCILLATION

Blocking anticyclones are, however, little more than a description of a particular pattern and do not explain why the strength of circulation, and with it the incidence of different patterns, change from year to year. About three-quarters of the time during northern hemisphere winter, there are just four or five broad types of circulation pattern. This suggests that there are tight constraints as

Right: A great ice bridge can form during the coldest winters at Niagara Falls, New York State, USA. This picture was taken during the unusually cold spell of the 1910s.

BLOCKING PATTERNS

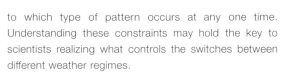

PATTERN A
When the NAO follows pattern A, there are strong westerly winds across the Atlantic, caused by a deep depression near Iceland and high pressure around the Azores. This pushes mild air across Europe and into Russia, while pulling cold air southwards over western Greenland. It also tends to bring mild winters to much of North America.

PATTERN B
When the NAO flips into pattern B, the winds follow a winding meandering pattern across the Atlantic Ocean. This causes a blocking anticyclone above Iceland or Scandinavia which pulls Arctic air down into Europe, with mild air being funnelled up towards Greenland.

CARL-GUSTAF ROSSBY (1898–1957)
Born in Sweden, Rossby first worked on meteorology with the Bergen School, before moving to the USA in 1926. His greatest contribution to understanding the climate was to provide a physical explanation for the large-scale wave-like motions in the upper atmosphere, which influence the movement of weather systems at lower levels. These waves are now known as Rossby waves. By explaining their behaviour in terms of the speed of the upper atmosphere westerly winds he provided forecasters with a powerful tool for predicting the motion of weather systems. He also made fundamental contributions to numerical modelling of the climate, which forms the basis of modern weather forecasting and the prediction of climate change.

to which type of pattern occurs at any one time. Understanding these constraints may hold the key to scientists realizing what controls the switches between different weather regimes.

The switch between blocking and symmetrical atmospheric patterns helps us to understand one of the strongest weather patterns, the North Atlantic Oscillation (NAO) – namely that, when winters are unusually severe in western Greenland, they are mild in northern Europe (pattern A) and vice versa (pattern B); see above. This see-saw behaviour was interpreted by Sir Gilbert Walker in the 1920s in terms of pressure differences between Iceland and southern Europe, and is seen as the northern cousin of El Niño in the Pacific.

The NAO may be caused by interactions that take place between the oceans and the atmosphere. The right combination of sea surface temperatures may dictate whether the NAO follows pattern A or B. For example, there is some evidence that the fluctuations taking place in the NAO are followed within a couple of years by shifts in the position of the Gulf Stream.

THE NAO AND GLOBAL WARMING
Since 1870, the NAO has fluctuated noticeably over periods ranging from several years to a few decades. It stuck to pattern A between 1900 and 1915, in the 1920s and, most notably, from 1988 to 1995. Conversely, it

followed pattern B in the 1940s and during the 1960s, bringing frequent severe winters to Europe but causing exceptionally mild weather in Greenland.

The NAO also has a significant influence on average temperatures in the northern hemisphere. If it follows pattern A for a long period, then the mild weather in Europe will be reflected in the annual figures. So, a significant part of the global warming since the mid-1980s has been caused by the very mild winters in the northern hemisphere. Indeed, since 1935 the NAO on its own can explain nearly one-third of the variability in winter temperatures for the latitudes 20–90°N. So understanding more about this major natural factor in climate variability is central to explaining the causes of global warming in the 20th century.

Left: Hot summer spells attract large numbers of people to the beach, as this overcrowded scene on the Baltic Coast of Germany shows.

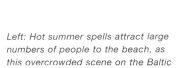

OCEAN CIRCULATION PATTERNS

OCEANS PLAY AN INTEGRAL PART IN TRANSPORTING
ENERGY AROUND THE WORLD – THE NORTH ATLANTIC,
IN PARTICULAR, IS PIVOTAL IN THIS PROCESS.

Contrary to popular opinion, the oceans are not fixed bodies of water but swirling transporters of energy, which have a key role to play in the climate.

THE GREAT OCEAN CONVEYOR

Surface currents and deep-water currents combine to drive the large-scale motion of the oceans around the world. The surface currents are caused by winds above them and the deep-water currents are driven by water that sinks and rises as a result of variations in temperature and salinity levels: this latter process is known as thermohaline circulation (see box, right).

The worldwide pattern of thermohaline circulation, the Great Ocean Conveyor (GOC), was first proposed by Wallace Broecker, at the Lamont Doherty Laboratory in 1985. The North Atlantic carries the majority of heat to the Arctic regions. Along the way the warm water gives up its heat through evaporation, making it increasingly cold and salty, and eventually enabling it to sink and form deep water that flows to the Antarctic. Here it is warmer and less dense than the locally frigid surface waters and so rises to become part of a strong vertical-circulation process. Descending cold water from around Antarctica flows northwards into the Pacific and Indian Oceans where there is no descending cold water. In the Atlantic this countercurrent is swallowed up by the much stronger southward flow.

PAST CHANGES IN THE GOC

Temperatures of countries at high latitudes are dependent on the temperature of the oceans that wash on their shores and so any change in the path of the GOC can have a huge impact on the climate of these regions. There is evidence that the path of the GOC can vary – this may be fundamental in defining past climatic change.

Analysis of the various layers in ocean sediments and ice cores that built up over the course of the last Ice age shows how suddenly some of the changes came about. In particular, the dramatic warmings that can be observed in the Greenland ice core appear to coincide with what are known as Heinrich layers in the ocean-sediment data. These layers, named after the scientist who first identified them, were produced by debris that was carried out into the North Atlantic Ocean by a huge surge of icebergs resulting from the sudden collapse of part of the ice sheet covering North America. These sudden influxes of freshwater appear to have altered the deep-water patterns of the North Atlantic by switching off the thermohaline circulation (see box, right).

FUTURE CHANGES IN THE GOC

There is no doubt that changing patterns of global ocean circulation help us to explain past climatic regimes. What is less clear is whether these sudden changes are relevant to the current debate about the impact of human activities. This revolves around the sensitivity of the GOC to changes in the amount of freshwater currently entering the northern North Atlantic Ocean. Modelling work suggests that circulation patterns are extremely sensitive to the run-off of rainwater from the continents, the number of icebergs calved off Greenland and the amount of precipitation from low-pressure systems tracking north-eastwards past Iceland and into the Norwegian Sea. Small variations in the total input may be able to trigger sudden switches in global ocean currents and bring about alternative patterns of circulation. If this happens, warm surface water would be carried less far north before sinking and returning southwards again. This would reduce sea surface

GREAT OCEAN CONVEYOR

Surface and deep-water circulation of the oceans is known as the Great Ocean Conveyor. Warm, low-salinity water from the tropical Pacific and Indian Oceans flows around South Africa and north to Iceland. In the North Atlantic it becomes more salty and cools. It sinks and flows back to Antarctica, where it combines with cold dense sinking water and flows north into the Pacific and Indian Oceans.

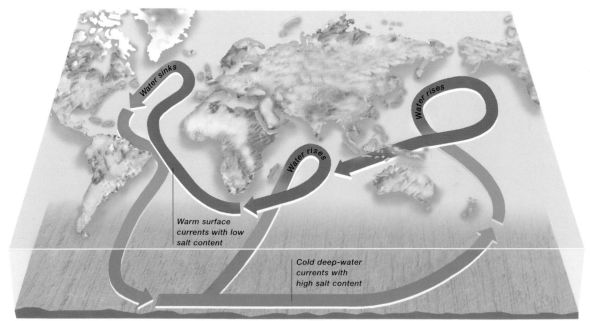

Warm surface
currents with low
salt content

Cold deep-water
currents with
high salt content

THERMOHALINE CIRCULATION

The density of sea water depends on its temperature and salinity: cold water is denser than warm water and salty water is denser than less salty water. But it is not a simple relationship. In terms of temperature, freshwater reaches its maximum density when it cools to 4°C (40°F) but, when water contains salt, its density can continue to increase at still lower temperatures. The densest water is found at –2°C (28°F) in the saltiest ocean water. In terms of salinity, the changes are simpler – the saltier the water, the more dense it becomes. Thus, when freshwater enters the oceans (from rivers, melting ice and rainfall, for example), it can float on the top of sea water preventing deep circulation. Similarly any surface warming that occurs from sunshine or the transport of warmer water from lower latitudes tends to form a stable surface layer. The stability of this surface layer of the oceans is controlled by the strength of the winds, which mix the water together. This in turn controls the depth of the shallow layer of water between the surface and deep waters – the thermocline. The mixing of water at very deep levels depends on the formation of high-salinity water, either by the process of freezing as salt is shed into the surrounding cold water or in areas of high evaporation. These processes are the principal mechanisms driving large-scale thermohaline circulation of the oceans.

Above: A false-colour image of the Gulf of Maine where upwelling cold water from the Labrador Current (dark blue and pink) meets much warmer water to the south of Cape Cod (yellow and orange), which is fed by the Gulf Stream.

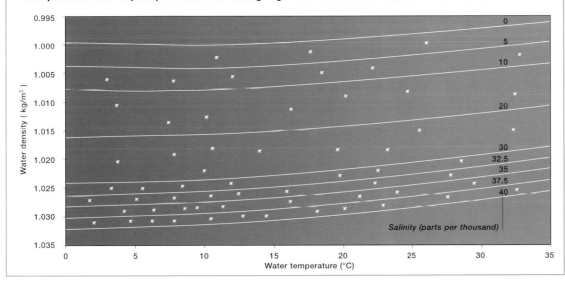

temperatures around southern Greenland and Iceland by 5°C (9°F) or more; it would also have a drastic impact on the climate of Europe and on the atmospheric circulation patterns of the northern hemisphere.

But, are the changes produced in the models by altering the balance between evaporation and precipitation in the northern North Atlantic realistic? These models can be tuned to produce chaotic shifts in circulation, but no such large switches have occurred in the last 10,000 years and

so this suggests that bigger perturbations are required to move the GOC into a different mode of circulation. Therefore, at the moment, the potential for our climate to undergo parallel shifts remains an unresolved issue. However, human activities are predicted to produce greater and more rapid changes than anything seen over the last 10,000 years and so maybe this is reason enough to worry about the future potential of the GOC to flip into a different mode.

Left: The ice-strewn waters around the South Shetland Islands in the Southern Ocean, where frigid waters sink to great depths as part of the process driving the Great Ocean Conveyor.

THE LITTLE ICE AGE

THE LITTLE ICE AGE IS OFTEN PORTRAYED AS A BITTERLY COLD PERIOD IN OUR CLIMATIC HISTORY WITH ICY WINTERS AND MISERABLE SUMMERS.

Above: Union Street, Ryde on the Isle of Wight, United Kingdom in January 1881. The island, which usually has a very mild climate, suffered unusually heavy snowfall.

During the Little Ice Age, which lasted from the 15th up to the mid-19th century, the weather was not consistently cold but very variable, with cold and warm periods striking intermittently.

THE TRADITIONAL PICTURE

The images of Brueghel's winter landscapes, Frost Fairs on the Thames and Dickensian Christmases are often presented as evidence that winters were much colder from the mid-16th century to the mid-19th century than they are today. Temperatures throughout the year were thought to be lower on the basis of crop failures, the expansion of glaciers and other historical records. This analysis was, until relatively recently, based largely on European observations. As more information about conditions around the world becomes available from proxy records, however, it is increasingly clear that this was not a period of wholescale global cooling, but rather a more complicated pattern of regional ups and downs.

EUROPEAN RECORDS

A revised picture of climate change during the last six centuries has been built up from a wide variety of sources. Written historical records clearly show that there were more cold winters in northern Europe during the Little Ice Age than there have been in recent decades. Records of

cereal- and wine-harvest dates show that the climate of the summers also had a tendency to be extremely variable. During the second half of the 16th century, the climate in Europe undoubtedly deteriorated. The glaciers in the Alps expanded dramatically and reached down into valleys by the end of the 1590s. In Switzerland, the period between 1570 and 1600 featured an exceptional number of cool wet summers and French wine-harvest dates provide confirmation of the poor summers that occurred during the late 16th century.

Moving on to the 17th century, evidence of the cold wet nature of the 1690s can be found in nearly all the available wine-harvest records from France and Germany, and from the Swiss summer records. The Swiss data also reveal the dramatic spring figures in the 1690s that show how the Alps suffered from particularly cold snowy conditions, which significantly delayed the growing season.

Subsequent records show how much the climate changed from decade to decade. The summers of the late 1770s, early 1780s and around 1800 were hot and yet those of the 1810s were cold and wet. The winters of the 1880s and 1890s were colder than previous decades and the 1880s and 1910s also had cool wet summers. Therefore, the striking feature was the marked oscillation between decades prior to the warming of the 20th century.

THE REST OF THE WORLD

A similar variable story emerges from other parts of the world. In eastern Asia, the 17th century was the coldest period, with several cold decades around 1800, but during the rest of the 19th century the region did not suffer the

Right: Hunters in the Snow (detail) by Pieter Brueghel the Elder, who, inspired by the Great Winter of 1565, established the genre of northern European winter landscapes which form an integral part of our image of the Little Ice Age.

sorts of low temperatures seen in Europe. Conversely, the North American records show that the coldest conditions occurred in the 19th century. Tree-ring data suggests that the 17th century was also cold in northern regions of the USA, but in the western states this period was warmer than the 20th century. Limited records for the southern hemisphere indicate that the coolest periods occurred earlier, mainly in the 16th and 17th centuries.

Recent studies covering a wide range of proxy data collected from around the northern hemisphere back up these earlier findings. They show that there was a cold period in the 1460s and confirm that there were lower temperatures at the end of the 16th century. Much of the second half of the 17th century and the 19th century were cold from around 1810 onwards. These cold conditions extended into the early 20th century.

The conclusion that can be drawn from all this evidence is that the Little Ice Age was not a monotonously cold period: certain intervals were colder than others. Only a few short cool episodes appear to have struck worldwide at the same time. These are the decades from the 1590s to the 1610s, the 1690s to the 1710s, the 1800s and 1810s, and the 1880s and the 1900s. Worldwide warm periods are less obvious, but the 1650s, 1730s, 1820s, 1930s and 1940s are the most striking. The lack of synchronicity means there is geographical variability in climatic anomalies, the coldest episodes in one region often not being coincidental with those in other regions.

20TH-CENTURY WARMING

What is beyond doubt is that, since the cool decades at the end of the 19th century, there has been a clear increase in the average global temperature of approximately 0.3–0.6°C (0.5–1°F). In the northern hemisphere the increase occurred principally during the 1920s and 1930s, and again since around 1980, with a pause, or even slight cooling, between the 1940s and the 1970s. In the southern hemisphere the warming trend has followed a steadier

pattern. Most striking of all is that the warmest eight years on record have been since 1982, with 1998 being by far the warmest year so far.

The regional distribution of the warming has exhibited some distinctive patterns, with the greatest warming occurring over the northern continents during the winter, especially across Siberia. In contrast, the North Atlantic Ocean south of Greenland has show a marked cooling since the 1950s and the central North Pacific has also been cooling in recent decades. These regional patterns are consistent with the fundamental role the mid-latitude westerly winds play in the natural variability of the climate.

Above: The Thaw, *painted by Claude Monet, is typical of the cold winters that were a common feature of the north European climate until the end of the 19th century.*

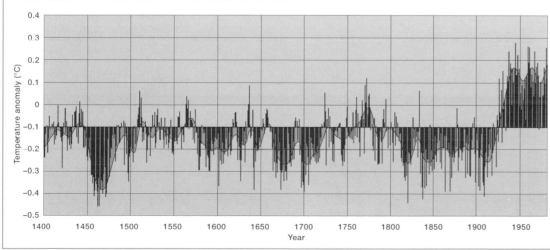

SIX CENTURIES OF NORTHERN HEMISPHERE TEMPERATURES
An estimate of annual fluctuations in the temperature of the northern hemisphere prior to the 20th century has been compiled from the combination of tree-ring, ice-core, historical and instrumental records. This shows that the climate warmed noticeably during the first half of the 20th century. Before that there were notable cool periods during the 1460s, around 1600, from 1660 to 1720 and from 1810 to 1920.

Global warming 184–5

THE PRAIRIES

CLIMATE OF PRAIRIES AND STEPPES

FROM CATASTROPHIC TORNADOES TO LONG PERIODS
OF DROUGHT, LIFE IN THE PRAIRIES POSES SOME OF
THE GREATEST CLIMATIC CHALLENGES.

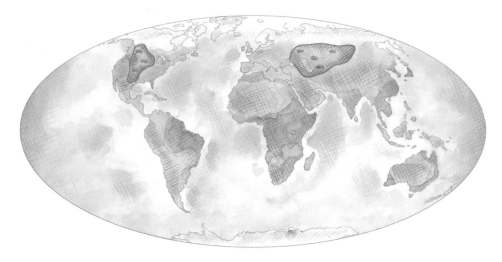

ALMA ATA, KAZAKHSTAN

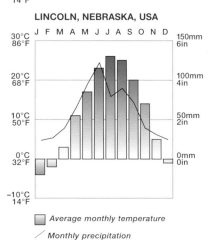

LINCOLN, NEBRASKA, USA

☐ *Average monthly temperature*

╱ *Monthly precipitation*

The great grassy plains found in the centre of the continents in the northern hemisphere have a climate of dramatic extremes.

A SEPARATE CLIMATIC ZONE?

The cool temperate grasslands of the northern hemisphere present something of a conundrum to biologists and climatologists. The vegetation of these regions is quite distinct and is defined by the extreme climate. However, it is difficult to explain exactly what makes their climate unique. Inspired by this challenge the biologist Köppen set out to publish a set of climatological zones in 1918 and it is still used by many geographers today. From study of rainfall and temperature data along the forest–steppe boundary, he explained why some areas are more productive than others in terms of agriculture. He constructed formulae that explain how effective rainfall is at different times of year and in different temperatures in producing fertile ground. Other biologists have since put forward various other formulae, but none is obviously a substantial improvement on Köppen's work.

It is clear that the broad strip of Asia from the east of the Black Sea as far as Mongolia, and in the broad triangle in North America jutting out in the lee of the Rocky Mountains as far as southern Lake Michigan, have a distinctive climate that warrants separate consideration. In North America the grassland triangle has an annual rainfall of 500mm (20in) or less because it is found in the rain shadow of the Rockies. The annual figures for the steppes of central Asia are roughly the same, not as a result of any mountain range but because of their distance from the sea.

It can be argued that the areas in the southern hemisphere on the Argentine Pampas, the High Veldt of South Africa and even the Canterbury Plains of New

Below: Wheat harvesting on the high plains of Montana, USA. In spite of the severe climate of the prairies, they are one of the world's most important grain-producing regions.

Zealand should be included in this climatic zone. It is, however, the extraordinary extremes and ferocious weather of the great grassland areas of the northern hemisphere that set them apart, and that have no southern equivalent.

TEMPERATURE EXTREMES

The grasslands of North America cover enormous expanses of land over a large range of latitudes, extending from northern Texas to Alberta, and so they experience markedly different temperatures from top to bottom. The essential feature is, however, the extraordinary temperature range between the summer and winter months, and from year to year, particularly with regard to the winter figures. For instance, in Havre, Montana, the lowest average January temperature is as low as –25°C (–13°F), while the highest is 1.1°C (34°F).

This variability is caused partly by the changing position of the storm tracks across North America from year to year, which bring different amounts of cold Arctic air and warm Pacific air to the region. The temperature fluctuations are also due to changes that occur when Pacific air crosses over the Western Cordillera mountain range. As this air rises it is kept relatively warm by the heat that is released through the condensation of water vapour, but, as it descends, the now dry air warms extremely rapidly and causes warm dry winds. Known as the Chinook in America and the Föhn in Europe where it sweeps down north of the Alps, this wind has a dramatic effect and consequences and is the stuff of legend. Stories from Alberta and Saskatchewan tell of how temperatures rose from –20°C (–4°F) to 4°C (39°F) in just five minutes when the Chinook swept through. The impact on snow cover is spectacular with thick cover disappearing in just a few hours as if by magic. These dramatic changes do not usually extend far beyond the shelter of the mountains and the warm air soon glides back up over the cold dense Arctic air.

In July the average temperatures range from 18 to 28°C (64 to 82°F). More striking are the daytime highs that range from around 28 to 34°C (82 to 93°F). The fluctuations from year to year are, however, much less dramatic than in winter.

THE NORTH AMERICAN MONSOON

Most of the rain falls on to the Great Plains of North America during the summer. For much of the grasslands triangle over 40 per cent of the annual total falls in June, July and August. This pattern, which has recently become known as the North American Monsoon, is caused by an upper-atmosphere high-pressure system that develops over northern Mexico and the desert south-west of the USA. This is analogous to the larger Tibetan High, which controls the monsoon over India. The south-westerly flow created by the upper-atmosphere high produces a sharp peak in the rainfall over western Arizona and New Mexico in July and August. The summertime rainfall peak extends through Colorado and into the Great Plains. At the same time the rainfall over the northern Rockies declines from June to August before picking up again in the autumn.

THE RUSSIAN STEPPES

The climatic patterns over the steppes of Russia are easier to describe than the prairies. The winter temperatures become more cold as you move further east. Although there can be dramatic swings in weather patterns, with substantial variations from year to year, there is no real equivalent to the extraordinary fluctuations induced by the Chinook. The average temperatures in July range from 17°C (63°F) to as high as 24°C (75°F) in the most southerly parts and show the same relative lack of variation as in North America. The rainfall is remarkably uniform, ranging from nearly 600mm (24in) in the west to around 450mm (18in) in the east. There is a pronounced summer peak in precipitation and this becomes more marked as you move to the east.

Above: The bleak landscape near Leninsk, Kazakhstan during the winter is typical of the steppes.

Thunderstorms 126–7

Tornadoes 128–9

The monsoon 138–9

THUNDERSTORMS

RUMBLING THUNDER, LASHING HAILSTONES AND
FLASHES OF LIGHTNING MAKE THUNDERSTORMS A
DRAMATIC AND SPECTACULAR CLIMATIC FEATURE.

Above: The threatening clouds of an
approaching spring storm over
Sundance, Wyoming, USA.

Thunderstorms can happen almost anywhere, but are most frequent over land in low and mid-latitudes during the hottest months of the year. A thunderstorm goes through a number of stages, from developing to mature and then decline, but normally lasts no more than half an hour.

WHY DO THUNDERSTORMS HAPPEN?

When the lower atmosphere is humid and unstable, conditions are ripe for a thunderstorm. The warm moist air at the surface starts to rise and, when it reaches the condensation point, clouds start to form. As the water vapour in the air condenses out into clouds, it releases latent heat, making the clouds rise still further. In severe storms, the rising air, an updraught, reaches heights of over 12km (7 miles) and travels at speeds of some 10m/s (33ft/s).

When it reaches the tropopause in the low levels of the upper atmosphere, the moisture-laden air stops rising and is spread out by the force of the jet stream to form the familiar anvil-shaped cloud – cumulonimbus. The cloud's rapid ascent means that the supercooled water droplets inside are swept up to high levels where, even though their temperature may fall to –40°C (–40°F), they may remain in a liquid form for some time or coalesce with ice crystals and snowflakes in the upper reaches of the cloud.

WHY DO CLOUDS RAIN AND SNOW?

A thunderstorm reaches its mature and most devastating stage when it starts to rain or snow. When the water droplets or ice crystals in the cloud become large enough and heavy enough to succumb to gravity, they fall to the ground as rain or snow. They create a downdraught of cool air as they fall, which counteracts the warm rising updraught. This happens because, when the falling water droplets and ice crystals reach warmer levels of the atmosphere nearer the ground, the ice starts to melt and absorbs some of the heat. This cooling process reinforces the downdraught (causing heavier rainfall) but stops the unstable updraughts of warm air, which initially set the storm off.

At this stage the storm goes into decline. However the downdraughts sometimes undercut nearby warm air, thereby triggering

Above left: A polarized-light photograph of a thin cross-section of a giant hailstone, which fell on Kansas, USA, in 1970. It shows how the hailstone was formed by a series of layers of ice.

Above: A giant irregularly shaped hailstone measuring some 15cm (6in) across. Hailstones this big can do immense damage and cause death and injury when they fall to Earth.

off another updraught and setting the next storm into action. This is why storms rumble across the countryside one after the other in a haphazard manner, each following slightly different tracks. It also explains how storms can strike a number of spots, while the areas in between remain dry.

SQUALL LINES

Not all thunderstorms are due to random unstable air masses. Others, often far more dangerous, develop in front of or behind a cold front, and are set off as the cold air sweeps over low-lying warm air. This makes them unstable as cold air is denser than warm air and tries to sink, a process that can fuel storms for longer, increasing their severity. A line of particularly intense storms is known as a squall line. The storms follow the path of air movement and are often heralded by a sharp line of threatening cloud and a sudden reversal of wind direction – a gust front.

HAIL

Hail is probably the most damaging form of precipitation in terms of potential adverse effects on agriculture. It can totally destroy fields of crops and can strip vines of their grapes. Hailstones occur in large thunderstorms, which have particularly strong updraughts. When raindrops start to fall, they are swept back up to high levels in these updraughts where they freeze. This process of ascent and descent repeats itself and the hailstone eventually builds up several layers of opaque and clear ice because it collides with supercooled water droplets at various levels in the cloud. The opaque layers form when the hailstone strikes droplets that are cold enough to freeze instantly, forming rime. Clear ice is formed where the cloud is warmer, but still below freezing, and the collisions form a liquid coating which then freezes to form a clear glaze.

When the hail becomes too large to be supported by the updraughts, it falls to the ground. Large hailstones are only formed when the hail rises and falls. This takes several hours and so only occurs in storms that have enough energy to keep the momentum going. In the most severe storms the hail often ranges from golf- to tennis-ball sizes. The record for the largest hailstone used to be held by one that fell on Kansas in 1970, weighing in at 0.76kg (1.67lb). But, this record was decisively broken in 1986 by the massive hailstone that crashed on to Bangladesh, weighing in at 1kg (2.2lb).

WEATHER RADAR

The detection, study and early prediction of severe thunderstorms have relied heavily on weather radar. This technology, similar to that used for air-traffic control, uses a transmitter to send out short, sharp pulses of microwaves and measures the signal reflected from the target. This can reveal a number of factors: the amount of signal reflected by different parts of the storm is a measure of how many droplets and ice particles are inside the cloud; the time taken for the signal to return indicates where the storm is situated; and, the shift of microwave frequencies emitted by the cloud reveals the wind speed within the storm.

Accurate interpretation of radar echoes is a complicated business because the amount of return signal depends on a variety of physical features of the thunderstorms, their precipitation patterns and the surrounding atmosphere. Nevertheless, over the last 40 years or so, atmospheric scientists have developed the technology and built up a comprehensive knowledge of the behaviour of storms. This has enabled them to predict how much rain will fall and to detect when the conditions within storms are ripe for the production of damaging hail, severe winds and tornadoes.

Above: Towering cumulonimbus clouds over Madagascar.

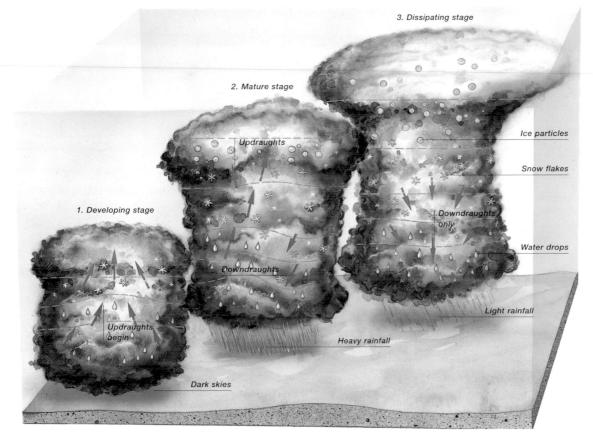

3. Dissipating stage

2. Mature stage

1. Developing stage

Updraughts

Downdraughts

Updraughts begin

Dark skies

Heavy rainfall

Ice particles

Snow flakes

Downdraughts only

Water drops

Light rainfall

THUNDERSTORMS AND HAIL

Warm moist air rises and, when it reaches the condensation point, clouds start to form. The air stops rising at the tropopause and the jet stream blows the clouds into an ominous-looking anvil shape. When the water droplets or ice crystals in the cloud eventually become large enough, they fall to the ground as rain or snow. The movement of the ice crystals and water droplets generates electic charges, which lead to lightning, and thunder is the sound that results from the violent heating of the atmosphere in the lightning strike. Hailstones occur in big thunderstorms – they are swept back up to high levels in the updraughts of rising air, where they freeze. As they ascend and descend in the clouds, they accumulate many layers of ice until they are heavy enough to fall to the ground.

Tornadoes 128–9

TORNADOES

DRAMATIC TWISTERS HAVE INSPIRED FILM-MAKERS AND AUTHORS ALIKE, BUT THEY REMAIN ONE OF THE MOST DESTRUCTIVE FEATURES OF THE EARTH'S CLIMATE.

WHERE DO TORNADOES STRIKE?
The areas of the USA most at risk from tornadoes vary according to the time of year. The most vulnerable area is across the Great Plains in late spring and early summer. Earlier in the spring, the greatest danger is in the south-eastern states. The regions shown in red in the map above are all vulnerable to tornadoes – those shown in deep red have suffered the greatest number over the last 40 years.

Tornadoes are the most devastating feature of severe thunderstorms and are responsible for hundreds of deaths every year. They can occur in many parts of the world, but are especially common in the Midwest USA.

WHAT CAUSES TORNADOES?
The simple answer to the question as to what causes tornadoes in the first place is that we do not know for sure. What we do know are the atmospheric conditions that are most likely to lead to an outbreak of tornadoes. They most frequently occur on the periphery of a severe thunderstorm, which is drawing in sufficient moist warm air to generate the rotational motion that sets off a tornado. In addition, this rising rotating air may interact with the horizontal wind field (surface wind) to produce more extensive rotation at low levels. Termed a mesocyclone, these complex rotating storm systems form a spinning column of air some 10–20km (6–12 miles) across.

WHERE DO TORNADOES STRIKE?
The prairies in the Midwest USA suffer more tornadoes than any other part of the world. This is due to the combination of warm moist low-level air from the Gulf of Mexico and cool dry upper air from the western USA, which together provide ripe conditions for generating the most severe thunderstorms. Tornadoes strike most frequently between the months of April and June, but they can occur at any time of year. Across the USA, they are most frequent in a corridor running from northern Texas up into Iowa which is often referred to as 'Tornado Alley'. They also happen in parts of northern India and Bangladesh, Japan, Australia, northern Argentina and northern Europe, including the British Isles.

Tornadoes are usually placed in one of three categories: weak, with winds of 64–180kph (40–112mph); strong, with winds of 181–331kph (113–206mph); and violent, with wind speeds in excess of 332kph (207mph). Around 80–90 per cent of tornadoes in the USA and the vast majority of those elsewhere in the world fall into the weak category. Strong tornadoes make up about 10–20 per cent of all tornadoes and account for many of the deaths recorded each year. Outside of the Great Plains of North America they are very rare, perhaps numbering five or six a year. Although only 10–20 violent tornadoes (1–2 per cent) are recorded each year in the USA, they cause a disproportionate amount of damage. In extreme cases they can have wind speeds of over 500kph (300mph), and sometimes the funnel may be as much as 1600m (5250ft) wide and extend to an altitude of 12km (40,000ft).

TORNADOES ON THE INCREASE
There is no doubt that the number of tornadoes observed in the USA has nearly doubled since the early 1950s. These figures do, however, depend on visual sightings, which mean that the categorization of the type of tornado witnessed is inevitably subjective. Much of the increase can be attributed to growing public awareness of the phenomenon and to increasing population in the areas where tornadoes occur, together with improvements in detection following the installation of weather radars. Therefore, we cannot be sure whether any of the increase is due to climatic factors, although it has been pointed out that the rise during the 1970s and 1980s was much smaller and so part of the upsurge in the 1990s may be the result of real climatic changes.

Right: A threatening thunderstorm, photographed in Kansas, USA, in springtime, with the funnel cloud of a tornado extending down from its base.

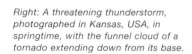

SEVERE STORM WARNINGS

The loss of life in tornadoes in the USA has been a major incentive for improving warning systems for severe storms. Many weak tornadoes are so ephemeral that predicting their touchdown is almost impossible. However, the most violent ones last for some time. The most severe tornado ever recorded struck in 1925 and travelled some 200km (125 miles) in three hours, crossing Missouri, Illinois and Indiana and killing 689 people.

Although radars cannot predict tornadoes very accurately, they can predict the conditions that are likely to generate them and short-term warnings are issued accordingly. These are an integral part of life for inhabitants of Tornado Alley in particular, and for many other parts of the USA in general. Although weather forecasts can identify the development of dangerous conditions many hours in advance, it is the use of Doppler radar that has transformed the production of emergency warnings because it can measure the intensity of rainfall and wind speeds. The most reliable indicator of dangerous conditions is a hook echo, which is clear evidence that a tornado is on its way. The US National Weather Service has invested some $2 billion in modernizing its warning system and installing 120 Doppler radars. This system is combined with upper-air data, surface observations and satellite imagery. Local forecasts can now issue 'watches' six to eight hours in advance.

Although the improved warning systems have reduced the number of fatalities, tornadoes still cause terrible local damage and loss of life. Since 1950, tornadoes have caused an average of 100 deaths per year – two-thirds of these

Left: A tornado that struck south of Inman, Kansas, USA in May 1999. The storm moved across central Kansas and set off other tornadoes that wreaked havoc further afield in Buhler.

were due to rare events. The most severe storms strike in April but the peak in twister activity is May and June, when about 30 per cent of the deaths occur. The south-east of the country has the highest proportion of the fatalities. This is partly because more violent tornadoes hit the south-east in spring than the less populated regions in Tornado Alley. The higher number of mobile homes in this region and the fact that many houses do not have basements because of the high water table adds to their vulnerability.

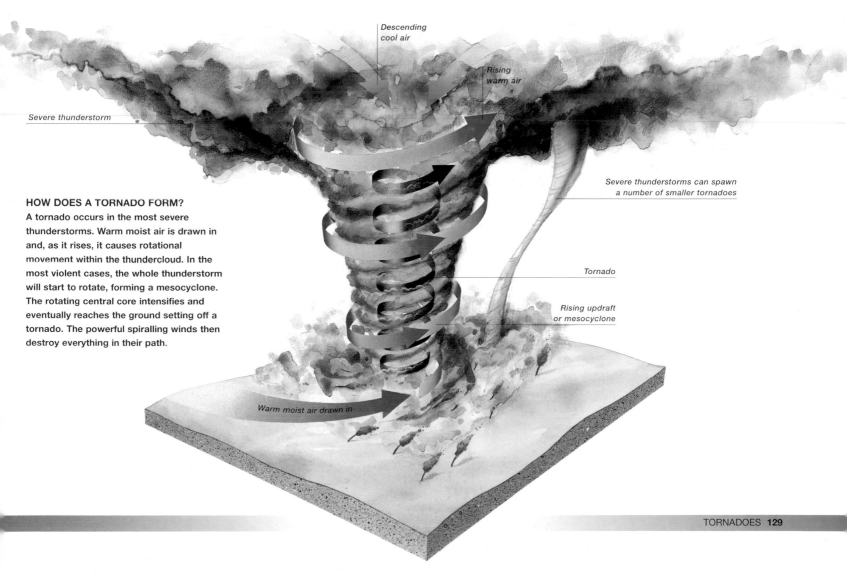

Descending cool air

Rising warm air

Severe thunderstorm

Severe thunderstorms can spawn a number of smaller tornadoes

Tornado

Rising updraft or mesocyclone

HOW DOES A TORNADO FORM?

A tornado occurs in the most severe thunderstorms. Warm moist air is drawn in and, as it rises, it causes rotational movement within the thundercloud. In the most violent cases, the whole thunderstorm will start to rotate, forming a mesocyclone. The rotating central core intensifies and eventually reaches the ground setting off a tornado. The powerful spiralling winds then destroy everything in their path.

Warm moist air drawn in

PLANT HARDINESS

FLORA THAT SURVIVE IN THE EXTREME CLIMATE OF THE PRAIRIES CAN TEACH GARDENERS THROUGHOUT THE WORLD ABOUT THE HARDINESS OF PLANTS.

Above: Mistletoe growing on a tree. The areas where this parasitic plant can grow in Europe are tightly defined by summer and winter temperatures.

The bitter winters and blistering summers of the prairies and adjacent parts of eastern North America have enabled botanists to identify the hardiness of native trees, shrubs and plants, as well as specimens imported to the region from around the world.

NATURAL RANGES

The vegetation that spreads across the continents of the northern hemisphere has adapted to major changes in the climate over the years. Many plants have also been introduced from other regions to meet various human needs and there is now an extraordinary mixture in the gardens and cities of Europe and North America. Vital to horticulture is an understanding of the source of these imports and the limits that they can stand. In this context, the plants in the eastern half of North America, which have to survive hot humid summers and severe winters, provide a useful measure of plant hardiness.

In other climatic zones the extent of native species provides more clues about the adaptability of plants. For example, in northern Europe species such as holly, ivy and mistletoe, which are symbolic of winter, have interesting climatic limitations. The northern limit of holly is defined by summer temperatures because it needs average July temperatures of at least 13°C (55°F). Its eastern limit is defined, however, by winter temperatures because it

cannot tolerate average temperatures much below 0°C (32°F) in January or February. Most of the hollies in Denmark were wiped out by the three consecutive cold winters from 1940 to 1942, showing that it is a species that is restricted to the milder parts of temperate regions. By comparison, ivy needs summer temperatures a degree or so higher but can tolerate winters that are that little bit colder. Mistletoe is much more continental, needing average summer temperatures of more than 16°C (61°F), similar to those experienced in southern England, but it can tolerate much harsher winters with average figures as low as –8°C (18°F), typical of western Russia.

Therefore, it pays to know where plants come from before making decisions on how to stock a garden because it can make all the difference between success and failure, not only with plants that come from extreme climates, but also with imports from warm parts of the world.

CHANGES IN PLANT DISTRIBUTION

The expansion and contraction of the ice sheets in the northern hemisphere over the course of the last million years or so have defined whether certain species survived in different parts of the world. In North America, where the most massive ice sheet developed, the plants were able to migrate north and south with the shifting climatic zones, but in Europe, the Alps acted as a mountainous barrier, which prevented species from migrating. As a result, species that are now common in China and the USA, such as the kiwi fruit, tulip tree and sweet gum, used to be present in Europe at one time, but they could not escape to the south of the Alps, and were driven to extinction by the successive waves of ice that struck. However, where they have been reintroduced they often thrive, as if they had never gone.

WHAT IS HARDINESS?

The distribution of many trees and shrubs in North America is partly defined by the low winter temperatures. This fact forms the basis of the widely used definition of hardiness zones for trees and shrubs. Devised by the Arnold

Right: The tulip tree. This native of the eastern USA used to grow in Europe until successive ice ages wiped it out. It has now been reintroduced and is grown successfully in many gardens.

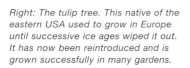

Arboretum of Harvard University, it links the hardiness of the plant with the average annual minimum temperatures (see below). In eastern North America these zones are easily defined in terms of geography, with the lowest temperatures occurring the furthest north. This means that only the Gulf Coast and Florida, zones 9 and 10, avoid severe frosts during the winter months of the year. Therefore, many of the plants from the eastern USA are hardy by European standards.

However, gardeners must be cautious in how they use the hardiness index for North American plants because it gives no information about the requirements of the growing season. For example, the glens on the Cairngorms in Scotland are classified as zone 6, but this does not mean that all the plants that grow from Cape Cod to El Paso (which is in the same plant-hardiness zone) will be able to survive in the Highlands. This region of the USA experiences blisteringly hot summers with heatwaves when temperatures do not fall below 25°C (77°F), and the daytime highs can be as much as 35°C (95°F) and higher for weeks on end. This bears no resemblance to the cool cloudy summers experienced in Scotland. Therefore, it is not surprising that species such as flowering dogwood, false acacia and stag's horn sumach – for all their hardiness – need the relative warmth of the south-eastern half of England in order to thrive.

Above: The prairies of northern Texas are covered with flowers in spring.

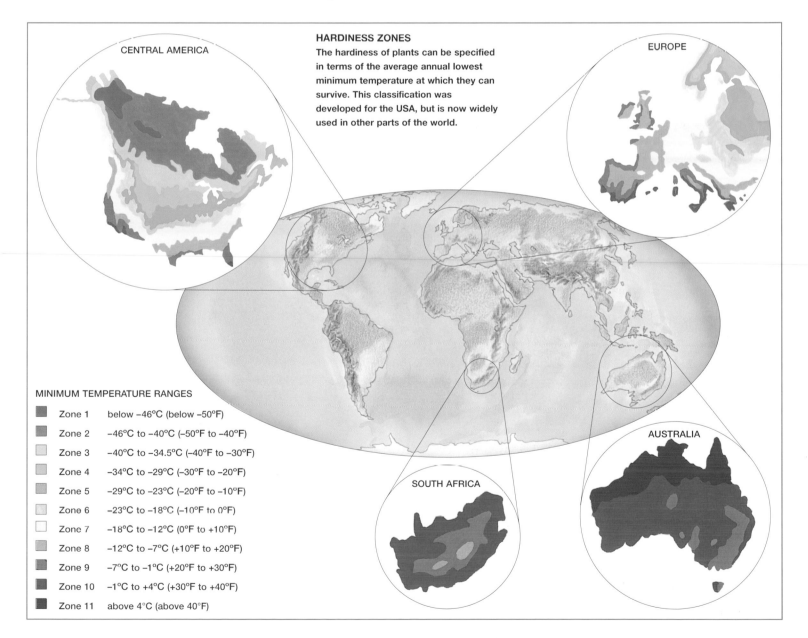

CENTRAL AMERICA

EUROPE

HARDINESS ZONES
The hardiness of plants can be specified in terms of the average annual lowest minimum temperature at which they can survive. This classification was developed for the USA, but is now widely used in other parts of the world.

AUSTRALIA

SOUTH AFRICA

MINIMUM TEMPERATURE RANGES

■	Zone 1	below –46°C (below –50°F)
■	Zone 2	–46°C to –40°C (–50°F to –40°F)
■	Zone 3	–40°C to –34.5°C (–40°F to –30°F)
■	Zone 4	–34°C to –29°C (–30°F to –20°F)
■	Zone 5	–29°C to –23°C (–20°F to –10°F)
■	Zone 6	–23°C to –18°C (–10°F to 0°F)
□	Zone 7	–18°C to –12°C (0°F to +10°F)
■	Zone 8	–12°C to –7°C (+10°F to +20°F)
■	Zone 9	–7°C to –1°C (+20°F to +30°F)
■	Zone 10	–1°C to +4°C (+30°F to +40°F)
■	Zone 11	above 4°C (above 40°F)

DUST BOWL YEARS

The extreme climate of the semi-arid grasslands of North America and Asia has exposed the weaknesses of some modern agricultural practices. The Dust Bowl years provide a good example of such bitter experiences.

The dramatic weather of the Great Plains of North America has always made farming and agriculture immensely demanding. The combination of bitterly cold, relatively dry winters and hot summers causes crops to lose a lot of moisture through evapotranspiration. This means that a successful harvest relies on summer rainfall, the amount of which tends to fluctuate significantly from year to year. Moreover, the hottest summers are the drought years so, when the rains fail, agriculture is in double jeopardy because the crops wither rapidly in the blazing heat.

Misery for farmers

Agriculture in these conditions has a history of setbacks that make it easy to underestimate the scale of the weather disaster that befell the region in the 1930s. The summers of 1934 and 1936 were by far the hottest since 1900, and 1936 was the driest by a large margin. Both were preceded by exceptionally dry springs. The ensuing disaster had a cataclysmic impact on farming over much of the Midwest USA.

The Dust Bowl years had a devastating effect on the region. First, there was an immediate drop in agricultural productivity – in 1934 and 1936 the average wheat yield across the Great Plains fell by about 29 per cent. Second, there was a mass exodus of settlers as people abandoned their unproductive farms. In the hardest-hit parts of Kansas and Oklahoma, more than half the population fled from their homes. This exodus was not as extreme as that of the 1890s, even though the earlier drought had not been as severe; this was because farmers at that time were even less well-equipped to handle drought and agriculture was not as important a part of the economy.

Political aid

This huge social disruption forced the Democrat Government to take action: it was an early example of major political intervention alleviating the impact of extreme weather events. The New Deal provided disaster aid, which enabled many people to survive the drought and set the precedent for similar action whenever severe weather strikes around the world. The fundamental lesson to emerge was that much of the agriculture that farmers were trying to produce on the Great Plains was not appropriate in such an arid region. In the wetter years of the 1920s farmers had extended their crops into areas with light sandy soil and low rainfall, which were not suitable for arable farming and should

Left: A sandstorm in Cimarron County, Oklahoma, USA shows the hardships that the local inhabitants had to face during the Dust Bowl Years.

have been left as pasture. Without grass to hold the soil in place, fierce winds and scorching heat stripped off huge quantities of topsoil.

In the long term, the Government purchased the unsuitable land, retired it from cultivation and seeded it with grass. This was combined with educational programmes that encouraged farmers to plant trees, provide shelter-belts, grow crops better-equipped for drought conditions, and introduce conservation methods (for example, contour ploughing, water conservation in irrigation ponds and strip ploughing to allow part of the land to lie fallow). Increasing rainfall meant that the lot of Midwest farmers improved into the 1940s. But, demands of food production during World War II meant that land unsuitable for agriculture was pressed back into service. When drought returned in the early 1950s many farmers had to relive past sufferings, which reinforced the case for the state and federal laws to protect the land from overexploitation. These experiences did not, however, deter the Soviet Union from doing the same thing in the steppes of Asia in the 1950s and 1960s. The opening up of the Virgin Lands was initially hailed a great success, but fell foul of the same problem of erratic rainfall and searing summers.

Measuring the impact

Tree-ring analysis confirms the enormity of the Dust Bowl years – no year back to 1700 matched the extreme drought of 1934, but there were periods during the mid-18th century when the region suffered sustained dry conditions comparable to 1934. Analysis suggests, too, that the level of rainfall in 1934 and 1936 is only likely to occur once every 250–400 years. There is also evidence of a 20-year cycle in the incidence of drought in the USA west of the Mississippi. Potentially more important is the recent analysis of lake sediments in North Dakota, which indicates that, prior to the year 1200, droughts in the Great Northern Plains were more frequent and intense, some lasting for centuries. It also presents striking evidence of a cycle of around 18.5 years.

An index for measuring drought was developed by Wayne C Palmer in the 1960s. Every week, it analyses the soil-moisture condition in the USA, using precipitation and temperature figures to compute evapotranspiration deficits or surpluses, which are compared with previous weeks and annual averages to derive a crop-moisture index. This is expressed in a scale from +4 or more (extremely moist) through zero (normal) to –4 or below (extreme drought). When calculated retrospectively, the maps for 1934 and 1936 show that the area of severe drought was far more extensive and prolonged than in any subsequent drought years, such as that of 1980.

The frequent duststorms of the 1930s – the terrifying 'Black Rollers' – made life well-nigh unbearable for those living in the Midwest. People could see the dust clouds approaching from many miles away like ominous thunderclouds, rolling along the ground and smothering everything in their path. They had to stay inside their houses and seal the cracks of doors and windows to stop the dust getting in. They tied handkerchiefs over their noses and mouths to protect their eyes from the stinging dust particles. The clouds of dust not only affected those living nearby but also blotted out the Sun in cities on the East Coast as far as 3000km (2000 miles) away. But many of those who were able to survive such harrowing events went on to thrive and today the region is as heavily populated as ever.

THE TROPICS

CLIMATE OF THE TROPICS

THE TROPICS ARE THE BOILER HOUSE OF THE EARTH,
FUELLING THE HEAT-EXCHANGE ENGINE THAT
CARRIES ENERGY FROM THE EQUATOR TO THE POLES.

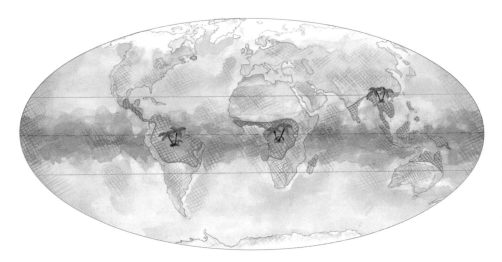

oceans air temperatures are governed by those of the sea surface, which mostly fall in the range 25–30°C (77–86°F). The rainfall over the oceans is more variable because it depends on wider climatic patterns.

DAWNING OF INTEREST

Much of the development of meteorology in the late 19th and first half of the 20th centuries centred on what was going on in mid-latitudes because this was the site of the action. It was also where most of the industrialized world was located, and so there were obvious incentives to getting to grips with weather events that could disrupt these countries. By comparison, the tropics looked far less interesting to meteorologists because, with the exception of the Indian monsoon, tropical storms and hurricanes, they seemed entirely bereft of exciting weather systems. The pressure pattern in equatorial regions fluctuated and in many places the main climatic variation was regular heavy showers in the late afternoon.

This blinkered view overlooked the fact that the tropics effectively drive the whole climate system because they act as a heat engine driven by the absorption of solar energy at low levels and low latitudes, which transfers heat to high altitudes and high latitudes. Several scientists working in the tropics, notably in India, where the mysterious

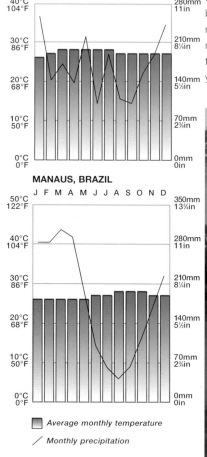

SINGAPORE

MANAUS, BRAZIL

■ *Average monthly temperature*

/ *Monthly precipitation*

Nearly 40 per cent of the Earth's surface lies between the Tropics of Cancer and Capricorn (23°N and S), but here we will concentrate on two main areas: the rainforests that cover the land and the oceans that occupy much of the tropics. Across the rainforests the temperatures average 25–30°C (77–86° F) throughout the year, and rainfall is high for much of the time. Over the

Below: Rainclouds over the Malaysian rainforest. The huge biodiversity of life found in the forests is due to plentiful rainfall and the near-constant heat and humidity in the region.

111 Pressure systems

Above: Dew on a leaf in the rainforest of Belize, Central America.

Left: The moist humid interior of this Venezuelan cloud forest shows the wide range of species that is able to thrive in the tropics.

variations in the seasonal monsoon played such a vital role in the lives of all the people living in the subcontinent, recognized that large-scale patterns were at work. But it was not until the 1960s, when the impact of events in the tropical Pacific became apparent around the world, that the importance of the tropics fully dawned on much of the meteorological community.

INTERTROPICAL CONVERGENCE ZONE

The area of intense convection and heavy rainfall that girdles the globe is known as the intertropical convergence zone (ITCZ) and its position fluctuates throughout the year. The movement to the north and south varies significantly from place to place, depending on the distribution of the continents and oceans. Fluctuations in how far the ITCZ extends to high latitudes has a profound effect on the rainfall in subtropical regions. It can also exert subtle influences on the circulation patterns that control the seasonal weather in mid-latitudes.

The movement of the ITCZ is controlled by a number of factors, including the annual march of the overhead Sun between the Tropics of Cancer and Capricorn and pulses of activity that run around the globe. Known as Madden–Julian Oscillations (MJO), after the scientists who first identified them, these consist of pulses of strong winds and rain that take 30–60 days to travel eastwards around the world. This phenomenon is at its strongest between the months of December and May, when it is one of the most intense weather systems in the tropics, pumping huge amounts of heat into the atmosphere. These bouts of activity have a noticeable impact on seasonal

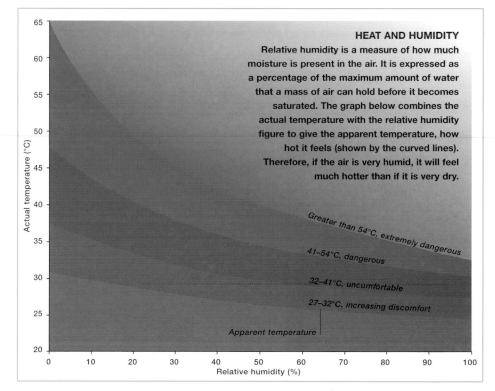

HEAT AND HUMIDITY
Relative humidity is a measure of how much moisture is present in the air. It is expressed as a percentage of the maximum amount of water that a mass of air can hold before it becomes saturated. The graph below combines the actual temperature with the relative humidity figure to give the apparent temperature, how hot it feels (shown by the curved lines). Therefore, if the air is very humid, it will feel much hotter than if it is very dry.

Greater than 54°C, extremely dangerous

41–54°C, dangerous

32–41°C, uncomfortable

27–32°C, increasing discomfort

Apparent temperature

Actual temperature (°C)

Relative humidity (%)

weather patterns, as well as on long-term developments in the tropics. The timing of their passage across the Indian Ocean affects the erratic development of the monsoon every summer. They can also trigger the development of an El Niño event if they happen to hit the western Pacific Ocean just as it is ready to hatch.

THE MONSOON

HALF OF THE WORLD'S POPULATION, LIVING IN THE TROPICS AND SUBTROPICS, RELIES ON THE SUMMER MONSOON FOR LIFE-GIVING RAIN.

Above: Torrential monsoon rain is responsible for flooding in the streets of Naogaon City, Bangladesh.

Much of the Indian subcontinent gets at least three-quarters of its rain in one intensely wet summer period. So if this rain fails to come one year, it can be a matter of life and death.

THE INDIAN MONSOON

The word monsoon is believed to come from the Arabic word 'mousim' meaning season. To meteorologists it has come to define prevailing winds that reverse to bring alternate wet and dry seasons. To the layman it is synonymous with a rainy season, and, in particular, the rains that come every summer to the Indian subcontinent.

The basic explanation for the monsoon was first proposed by Sir Edmond Halley (see box, far right) to the Royal Society in 1686. He suggested that as Asia warms up during the summer months it draws huge quantities of moisture northwards from the tropical Indian Ocean, which then fall as torrential rain. In winter the reverse occurs with the winds blowing from the cold continent towards the warm sea, resulting in dry weather. This simple explanation is only part of the story. The summer monsoon is so much stronger over India than elsewhere because of the controlling influence of the Himalayas and the Tibetan plateau. What seems to happen is that, as the Indian subcontinent and the rest of Asia heat up, the jet stream in the upper atmosphere, which guides weather systems, switches from the south of the Himalayas to the north of Tibet. This is a big shift and so moist tropical air, and with

it heavy rain, is drawn further north over India than it is in other parts of the subtropics. But it is the timing of the switch that is crucial to the strength of the monsoon.

The yearly waxing and waning of the monsoon across India seems inexorable. It usually bursts on to the Malabar coast in the south-west at the end of May and sweeps up across the country to reach New Delhi by the beginning of July. During September, as Asia cools, it starts to retreat, leaving the south of the country at the end of November.

VARIATIONS FROM YEAR TO YEAR

Every year the monsoon movement takes place in fits and starts. Periods of torrential storms are followed by days of sunshine. More importantly from year to year the timing of the start and end of the monsoon varies greatly and it can sometimes be very short. Unpredictable rainfall can be disastrous for agriculture. The spectre of famine has haunted India for centuries and the monsoon used to be less reliable before 1920. So although the rain has been plentiful since then, we cannot bank on the fact that it will carry on that way.

Although the subject of intense study, the underlying causes of the wayward behaviour of the monsoon are not yet fully understood. Indeed, the monsoon was the subject of the earliest scientific efforts at long-range weather forecasting after the terrible famine of 1878. The Government of India called upon H F Blandford, the Chief Reporter of the Indian Meteorological Department, to prepare monsoon forecasts. Blandford concluded that 'the varying extent and thickness of Himalayan snows exercise a great and prolonged influence on the climate conditions and weather of the plains of north-west India'. This relationship suggested that, when there was deep snow cover during the spring there would be a dry summer and, when there

MONSOONS AROUND THE WORLD

Changes in temperature of the land masses of Eurasia and Africa cause a dramatic shift in the rainfall between winter and summer. Half the world's population lives under the influence of these seasonal rainfall patterns.

WINTER MONSOON

During the winter half of the year cool dry air flows south-westwards from Asia over the Indian subcontinent. This brings dry sunny conditions to most of the region for the period from October to April, although the dry conditions do not extend to southern India and Sri Lanka until the end of the year.

Winds blow off India onto the sea

ITCZ remains close to equator

Equator

SUMMER MONSOON

As the temperature rises over Tibet during the spring and summer, warm moist air is drawn up from the Indian Ocean across the Indian subcontinent bringing spells of heavy rainfall. These start in the south in late May and usually move up across the country during the next two months. They then start slowly to recede back south-westwards in September.

Winds blow off the sea onto India

ITCZ moves northwards bringing torrential rain to India

Equator

Over the last 130 years the general trend for the average seasonal rainfall over the whole of India during the monsoon has remained virtually constant. There have, however, been noticeable changes over shorter timescales, with greater fluctuations from year to year before 1920, and more dry years between the mid-1960s and the late 1980s.

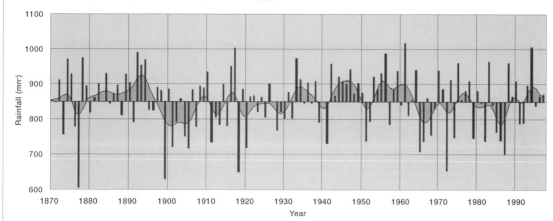

EDMOND HALLEY (1656–1742)
Halley was an astronomer and mathematician who made a number of significant contributions to science. He is best known for his work on celestial mechanics including his calculation of the orbits of 24 comets, which enabled him to predict the return, every 76 years, of the comet that bears his name. His contributions to meteorology included papers on the monsoon and the trade winds, and he established the mathematical law connecting air pressure with height above sea level.

was only a thin cover of snow, there would be plentiful rain. Initial forecasts between 1882 and 1885 were relatively accurate and encouraged Blandford to start operational forecasts covering the whole of India and Burma by 1886. These warned local inhabitants whether they should expect a good year or a bad year and whether they needed to start saving water. Since then, predicting the monsoon has been an important task for the Indian Meteorological Department.

ACCURACY OF FORECASTING

Predicting how the erratic movement of the jet streams will affect when the monsoon strikes India depends on a number of meteorological factors. Scientists established that El Niño was significant in the development of the monsoon, but between the 1920s and the early 1980s the forecasting methods made little progress. Although new climatic links were identified and some of the established relationships fell from favour, such as the Himalayan snow-cover link, which was dropped from forecasts in the 1950s, the general forecast performance stagnated.

Studies since the early 1980s have shown that the relationship over time between the various meteorological predictors around the world and the subsequent monsoon is important in developing improved forecasts. However, one problem has been the quality of the data. For example, recent satellite observations have revived interest in the level of Himalayan snow cover because the extent of Eurasian–Himalayan winter snow cover was proving to be useful. In addition, the role of sea surface temperatures in the Pacific has been established more clearly: warm El Niño events precede dry summers and cold La Niña events, when the eastern equatorial Pacific experiences temperatures well below normal, produce plentiful rainfall.

The new forecasts rely on 19 significant meteorological factors, many of them closely interrelated, including various local weather conditions, northern hemisphere winter temperatures and the Quasi-biennial Oscillation. Selections of these predictors can deliver impressive forecast records, but even the best combination of data varies from time to time and so leads to inaccuracies in forecasts. For instance, although El Niño events coincide with dry years in India, there have been periods when this has not been the case: the prolonged El Niño that occurred from 1991–5 did not cause a severe drought in India. In addition, Madden–Julian Oscillations (the pulses of winds and rains that affect the position of the ITCZ) appear to interact with the development of the monsoon, in a similar manner as they do with El Niño, beefing it up or slowing it down depending on what time of year it is.

Despite these challenges, predicting the monsoon more accurately is immensely important to India: the dependence of the country on this well-defined but little understood seasonal weather movement makes the monsoon possibly the single most important challenge for forecasters. Until that challenge is met the monsoon will, in Thomas Carlyle's words, reflect the 'mysterious course of Providence'.

Left: This image, taken by NOAA-14 satellite, of the Indian subcontinent during the monsoon season shows a depression and cloud cover that are typical for the time of year.

LIFECYCLE OF A HURRICANE
Conditions ripe for hurricane formation occur when strong convection, generated by warm moist air rising over a sea surface with a temperature above 27°C (81°F), combines with low-level winds. The pressure starts to fall at the centre of the system and, over a few days, the winds rise in a tight band of convection of some 30–60km (20–35 miles) radius and the clouds gust in a remarkably circular pattern around a central 'eye' of a mature hurricane. This lasts on average for two or three days, or until the storm hits land or moves over cooler water.

Rising air emerging from the top of the hurricane spreads out at high levels in a clockwise pattern

Warm moist air rises

Descending dry air in the calm 'eye'

Sea surface temperature above 27°C (81°F)

Low-level winds

Low-level winds swirl inwards in an anticlockwise direction

TROPICAL STORMS AND HURRICANES

The tropics spawn the most formidable storms that take place in the world. These powerful systems exert a major influence on the climate both in low latitudes and further afield.

Known as hurricanes in the Atlantic, typhoons in the north-western Pacific and cyclones elsewhere, tropical storms are fuelled by the oceans from below and steered by winds at high levels, making them a threat to many parts of the world and a challenge to forecasters.

Path of a hurricane
Tropical storms start life as minor depressions that are born over the oceans along the edge of the ITCZ. They only develop over water with a temperature above 27°C (81°F) and their intensity is dictated by sea surface temperatures. These systems, common features of tropical weather, frequently produce squally weather and heavy rain. Many produce little more than weak convection but some develop into major storms. The circular cloud system (see left) expands as the storm matures. The maximum wind speed around a storm depends on how organized and wide the circular area is. As the storm grows, it moves westwards in the trade winds, usually in the belt 8–15°N and S, and veers towards higher latitudes.

Some tropical cyclones grow and develop a radius of more than 300km (186 miles) before they start to decay. The declining phase is accelerated when the hurricane passes over land or cold water. By this stage the hurricane has reached higher latitudes and is being pushed eastwards by the mid-latitude westerlies. At higher latitudes, hurricanes sometimes have one more trick up their sleeve. They can pick up a new lease of life with great rapidity by becoming reinvigorated by transferring energy from a nearby weather system to generate an explosive new extratropical depression.

Dangers at sea
At sea, hurricanes can be dangerous because they cause high winds and waves. The highest waves occur on the right of the hurricane's path. These waves and a strong swell travel ahead of the storm and provide warning that it is on its way. As the storm comes near to the shore, the waves pile up ahead of the hurricane and, when the winds are particularly strong, a storm surge – a rise in sea level – forms. This wall of water can do the most damage when hurricanes strike the shoreline, especially if it coincides with a high tide.

Classifying hurricanes
The most widely used classification of tropical cyclones is the Saffir–Simpson table, which grades hurricanes on a scale of one (the weakest) to five (the most devastating), linking central pressure, wind speed and storm-surge height. Strictly speaking this scale applies only to hurricanes in the Atlantic basin, because the properties of tropical cyclones in other parts of the world are slightly different (for example, the central pressure of storms in the north-west Pacific tends to be 6 millibars lower even though there are comparable wind speeds). Furthermore, because many of these storms develop in remote areas, they are often classified in terms of what they look like in satellite images. This technique focuses on how symmetrical the storm is: the more symmetrical it is, the more intense it is. Known as the Dvorak scale, it is widely used by tropical-analysis centres around the world to measure the intensity of storms.

Predicting hurricanes
Meteorologists are now able to make broad empirical forecasts (based on observations rather than climatic models, which produce standard numerical forecasts) about the level of hurricane activity in the tropical Atlantic by relying on forecasts of El Niño as well as on other climatic factors. For example, when there are high sea surface temperatures in the eastern Pacific, hurricane activity is low and when the temperature falls below normal the activity peps up. Other factors included in these forecasts are rainfall levels in the Sahel, sea-surface temperature in the tropical and north Atlantic and the associated atmospheric-pressure patterns, and the direction of the Quasi-biennial Oscillation (QBO).

These forecasts give estimates of the number and intensity of tropical storms in the season ahead and offer guidance on which areas are most at risk. Their track record up to 1997 was good, but the rapid El Niño warming in that year caught forecasters out: it produced a quiet hurricane season and the rapid shift to La Niña in 1998 led to unexpectedly high activity. Nevertheless, these empirical forecasts are useful for anticipating seasonal behaviour. The prediction of the movement and intensity of individual storms up to about five days ahead remains the preserve of the standard numerical forecasts.

Improved forecasting and monitoring of hurricanes has reduced fatalities, especially in the USA. Once a major storm has been identified the first priority is to predict where it will strike land and to alert the emergency services so that they can move people at risk inland or to higher ground. Although there is still some uncertainty about the track hurricanes will follow, these warnings have reduced the death toll in the USA dramatically. In Bangladesh, where the huge expanse of low-lying ground makes it impossible to evacuate areas at risk, volunteers guide people to specially constructed mounds and cyclone shelters. This method was effective in the 1980s, but the death of some 130,000 people in 1991 shows how vulnerable the country remains to cyclones whirling in from the Bay of Bengal.

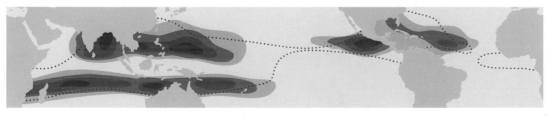

DISTRIBUTION OF HURRICANES
This map shows where hurricanes are most likely to occur. The areas where they start are shown in red – areas of dark red indicate the highest incidence of formation. The dotted line shows where the sea surface temperature is greater than 27°C (81°F) in the warmest month.

THE BIODIVERSITY OF LIFE

AN EXTRAORDINARY WEALTH OF SPECIES LIVES IN THE
TROPICAL RAINFORESTS BECAUSE THE STABLE CLIMATE
FOUND THERE IS IDEAL FOR SUSTAINING LIFE.

The immense diversity of life in the tropics is now under great threat from human overexploitation and an international effort is needed to ensure its survival.

TEEMING WITH LIFE

We do not know exactly how many different species live on the Earth. The total could range from 5 to 50 million, but it is estimated that the tropical rainforests harbour over half of this total even though they make up a mere 6 per cent of the land surface. At a local level the numbers are more staggering. About 300 tree species were found in 1 hectare (2.5 acre) plots near Iquitos, Peru, and over 1000 species were found in 10 selected plots of the same size in Borneo. This compares with 700 native tree species found in the whole of North America. The equivalent butterfly and beetle figures are even more startling and confirm not only the immense diversity, but also the fact that a high proportion of species surveyed was previously unknown to science.

This lack of knowledge may have vital implications for medicine. For example, there may be several hundred thousand different types of plant in the rainforest, some of which could contain substances with powerful healing properties. When the American National Cancer Institute

Right: A view of the rainforest canopy in Venezuela, where a wealth of different plant and animal species are found.

screened 12,000 plants, they identified three species with highly significant anti-cancer properties: a success rate of one in about 4000 species in this one disease category alone.

A HOTHOUSE ENVIRONMENT

The richness in diversity of life forms came about because of forest conditions: near-constant heat, humidity and heavy rainfall. The average temperature is 27°C (81°F) and it rarely ever rises above 35°C (95°F) or falls below 20°C (68°F). The rainfall exceeds 2000mm (80 inches) a year and is spread throughout the seasons. When combined with plentiful sunshine these conditions are ideal for plant growth.

The climate of the area has been stable for tens of thousands of years. Although there were some drier periods, fluctuations from year to year were small compared to those at higher latitudes. Moreover, conditions did not seem to change noticeably during the ice ages and so served as a sanctuary for many forms of life when the rest of the Earth was experiencing dramatic climatic variations. The climate stability is reflected in the longevity of many of the trees – the largest in the Amazon are often 500–1000 years old.

EACH TO ITS OWN NICHE

The degree of specialization is extraordinary. Many creatures exploit the tiny range of environmental opportunities provided by epiphytes (plants that grow on trees but do not extract water or nutrients from them). So, for example, an exquisite damsel fly or a particular variety of brilliantly coloured poison-dart frog may become wholly dependent on living in such plants, using the constant supply of rainwater that collects in the leaves to raise their young. They do not need to be able to migrate to avoid harsh winters or searing drought and in their cloistered little world all their needs are met by the single plant. By way of contrast their opposite numbers in Europe or North America have had to adapt to a far wider range of conditions and a much smaller number of species has survived this climatically dominated evolutionary process.

OVEREXPLOITATION OF RAINFORESTS

One of today's most pressing environmental issues is the exploitation of the tropical rainforests for agriculture and timber, which is destroying this treasure house of biodiversity. The fragmentation of the forest and creation of cleared areas alter the local climate and threaten the conditions needed to maintain many species. Large-scale destruction such as the fires in Indonesia resulting from an El Niño-related drought in 1997–8 only serves to accelerate this process.

The way to protect this diversity is to exploit its commercial potential by, say, developing new medicines. For example, the Government of Costa Rica and a pharmaceutical company have made an innovative deal in which Costa Rica conserves an area of forest, supported by a payment in return for access to the results of searching for plants that contain medically valuable products in the forest. The company will then pay Costa Rica a royalty on products developed from the research. The deal represents the first step in providing a conservation agency in a developing country with a financial stake in the intellectual property of its biodiversity.

Above: The red-eyed tree frog, which lives in the cloud forests of Costa Rica. Recent warming of the tropical Pacific has reduced the incidence of cloud here and threatened the survival of many species of frog.

Left: The buttress roots of the Blue Quangdong tree in the rainforest of the Mount Spec National Park, Queensland, Australia.

Destruction of the rainforests 144–5

El Niño 146–7

DESTRUCTION OF THE RAINFORESTS

THE DESTRUCTION OF THE RAINFORESTS HAS BEEN
HEADLINE NEWS FOR YEARS, BUT WE ARE STILL UNSURE
OF WHAT THE LONG-TERM CONSEQUENCES WILL BE.

Deforestation of the tropics is one of the most powerful images of the impact of human activities on the environment. The climatic consequences of this destruction are more complicated than might seem likely at first sight.

THE SCALE OF DESTRUCTION

Most of the remaining tropical rainforests are in the Amazon basin, which contains about half the world's total, with the rest of Latin America containing another 10 per cent or so. Next comes Asia with nearly a quarter of the forests, concentrated mainly in Indonesia, Malaysia, the Philippines and Indochina, and the remainder are found in equatorial Africa, principally in the Congo and Zaire.

The scale of destruction has been greatest in Amazonia but all these regions have suffered substantial clearing in recent years. When it comes to giving numbers, it is difficult to be precise. Measurements from space provide an indication of the scale of clearing, but cannot provide absolute figures, often because it is hard to know exactly what to measure because the reasons for the destruction of the rainforest differ from place to place. In some cases, industries require particular trees and so parts of the forest are left intact, while in other cases entire patches of land are cleared and replaced with agriculture.

On a planetary scale, these forest are important to the atmosphere because, when they grow, they take carbon out of the atmosphere and store it. When they are cut down and burnt, the stored carbon is released into the atmosphere as carbon dioxide. Estimates suggest that the amount of rainforest destruction that has already taken place has increased carbon dioxide levels in the atmosphere by 5–20 per cent.

MODELLING THE CLIMATIC EFFECTS

Global computer models – GCMs – are used to predict the consequences of deforestation. The replacement of the rainforests with grassland will increase the planet's albedo

Right: A false-colour Landsat image of deforestation in Brazil. The colours have been chosen to highlight the destruction of the rainforest: the dark green of the natural forest contrasts with the pale green and pinks of the areas that have been cleared. Latest estimates based on Landsat images suggest that this deforestation is damaging some 2.5–3.5 million hectares (6.4–9 million acres) of forest in the Amazon basin each year.

because more sunlight will be reflected back into space by the altered vegetation. The GCMs must take into account the dramatic changes that will also occur to the water cycle as a result of the change in vegetation cover. This, in turn, will alter the amount of evapotranspiration, cloudiness, rainfall and run-off: all of which have far-reaching climatic implications. In addition, the change in surface roughness in going from tall forest to grass and low scrub will alter wind speeds, and this will have an influence on the surface temperature and hence the amount of heat radiation emitted by the surface.

In addressing these various changes, different modelling exercises have made different assumptions about how certain surface characteristics will respond to deforestation. Although these figures are based on experimental results from measurements obtained in the field, the range of figures leads to completely different estimates of what the regional and global consequences of widespread deforestation could potentially be. In one extreme model, when the Amazon basin is deforested, the temperature rises by up to 2.5°C (4.5°F), the annual rainfall of approximately 2000mm (80 in) falls by about 800mm (32in) and the evapotranspiration by a little over half this figure. However, an alternative simulation, with a different set of

assumptions, showed that there was little change in surface temperature and only a 10 per cent reduction in precipitation and evaporation levels. The moisture changes were a regional phenomenon and no major global-scale changes were detected.

The importance of these studies is not whether one or other of them is 'correct', but to show just how complicated the consequences of changing one part of the global climate system are. We return to this issue in the last chapter of the book when we consider more aspects of the impact of human activities on the climate.

NATURAL DISASTERS
While deforestation in the tropics is principally caused by human intervention, natural variability of the climate can also play a part. During the 1982–3 El Niño event, the drought over Borneo led to some 3–4 million hectares (7.5–10 million acres) of forest being destroyed by fire. The intense El Niño in 1997 8 produced the same problems. Although many of the fires were started deliberately by those wishing to exploit the conditions to produce agricultural land, the fluctuation of the climate made it much easier to start the fires, and much more difficult to put them out when situation began to get out of hand.

Above: Rainforest in the Petan Jungle, Guatemala, being burnt to clear land for agriculture. This method of land clearance is quick and simple, and the ash returns nutrients to the soil. But heavy rainfall and thin soil mean that the nutrients are soon leached away and productivity drops, so the farmers often move on after a few years.

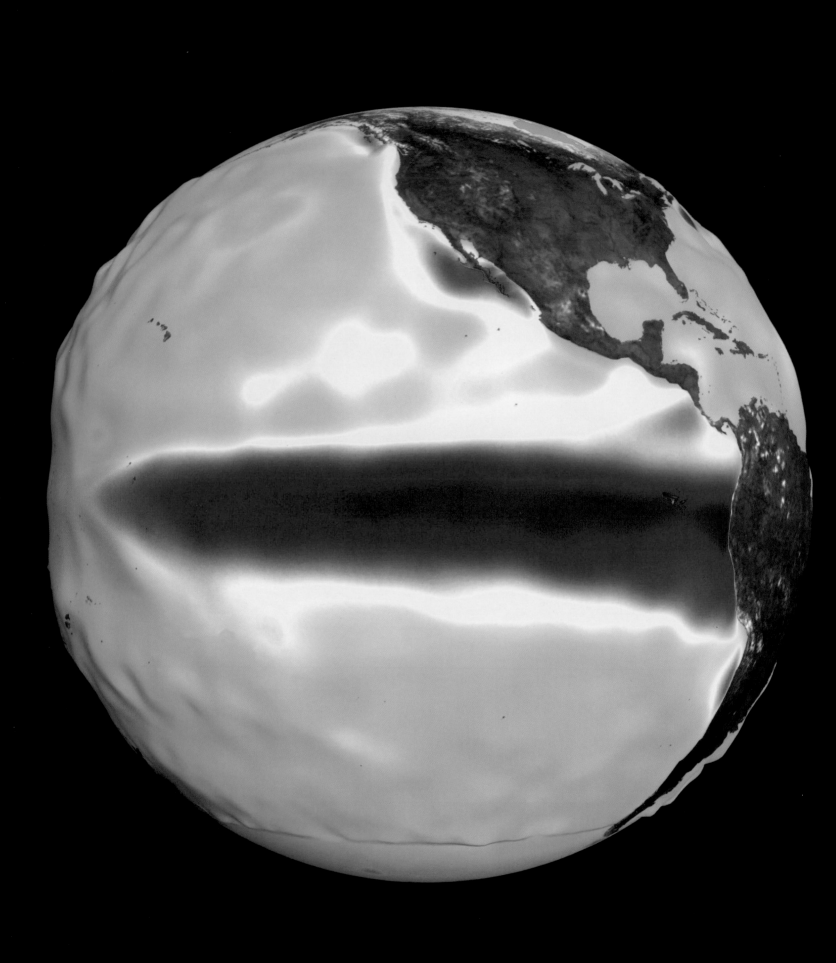

EL NIÑO

An El Niño event is the reversal of the usual weather patterns that occur in the tropical Pacific, caused by irregular fluctuations in sea surface temperatures. It can bring drought to Zimbabwe and floods to Kenya, forest fires to Indonesia and rain to the parched Levant.

Peruvian fishermen discovered centuries ago that in some years warm water tended to spread southwards from the equator around December time. This warm layer of water trapped cold upwelling nutrient-rich water that sustained abundant stocks of fish and, when these warm events were particularly strong, they had a catastrophic impact on fishing. El Niño derives its name from the Spanish for 'little boy' because its timing meant it was linked with the Nativity. We now know that these events form part of widespread atmosphere–ocean interactions.

Normal conditions in the Pacific

In normal conditions, equatorial winds blow from South America towards New Guinea because the atmospheric pressure is higher in the eastern and central Pacific and lower in the west. These winds push warm surface ocean currents westwards, bringing warmer water and higher sea levels to the west. This means that the sea that washes up on the coast of New Guinea is warmer in temperature and at higher levels than that on the coast of South America.

As the warm surface water is dragged away by the winds, upwelling cold water rises to replace it in the east Pacific, which brings the thermocline (the boundary separating the warm upper ocean from the colder deep water) closer to the surface. This means that sea surface temperatures are low close to the equator, typically 20–24°C (68–75°F), and a long tongue of cool water extends westwards towards the International Dateline. This cool water keeps air temperatures down and suppresses convection in the atmosphere. In contrast, in the west of the Pacific the sea surface temperatures are high, typically 28–30°C (82–86°F), and this generates a region of strong atmospheric convection. This pattern establishes a convective loop of warm air rising in the west, flowing eastwards and sinking in the east, which reinforces the climatic pattern of the region.

What is an El Niño event?

In an El Niño event, the winds change direction as part of a wider periodic shift of atmospheric-pressure patterns over the Indian and Pacific Oceans, which has been known as the Southern Oscillation since the 1920s. Warm surface winds in the western tropical Pacific flow eastwards, and the eastern and central tropical Pacific

Left: A satellite image of the Pacific ocean in December 1997 during an El Niño event. It shows sea surface height and temperature: red indicates areas that are warmer than normal.

becomes warmer and wetter than it is under normal conditions. Warmer water extends eastwards, pushing the cold water to deeper levels and depressing the thermocline. The area of strong convection expands towards South America and this establishes a reverse atmospheric-circulation pattern to normal, which maintains El Niño conditions. The close link between the El Niño and the Southern Oscillation means that meteorologists sometimes talk about El Niño Southern Oscillation (ENSO) to describe the whole phenomenon.

What causes El Niño?

For climatologists the challenge of ENSO is to explain what causes the switch from El Niño conditions to the opposite (La Niña), rather than one or other condition remaining in place forever. Models that show how Pacific winds and currents can swing back and forth every few years are the key both to understanding the processes and to producing better forecasts of how El Niño will develop a year or more ahead.

These changes in circulation are somehow linked to the slow-moving undulations in the thickness of the ocean surface layer, which slosh back and forth across the tropical Pacific. There are two types of ocean wave that cause this change in the depth of the thermocline. Kelvin waves move eastwards close to the equator and take two to three months to cross the Pacific. Rossby waves move westwards, taking nine months to cross the Pacific over the equator but much longer over higher latitudes because they are controlled by the circulation of the globe (for example, at 12°N and S they take four years to cross the Pacific). When both these types of waves reach the edges of the Pacific they tend to be reflected back in the direction they have come from, but they switch type, so Kelvin waves return as Rossby waves and vice versa. Computer models of these processes realistically reproduce atmospheric-circulation patterns, which switch back and forth between El Niño and La Niña conditions every three to five years.

Since the record-breaking El Niño event of 1982–3, the fluctuations in the Pacific conditions seem to have become more dramatic. There was a significant warming in 1987–8, followed by a marked La Niña a year later. Much of the period from 1991–5 saw a prolonged El Niño event, which switched to La Niña in 1996. Then in early to mid-1997, contrary to most predictions, there was a sudden warming to levels that matched or exceeded the 1982–3 event, only to switch suddenly into La Niña in 1998. These unexpected events show how difficult it is to produce forecasts for the equatorial Pacific.

The disastrous harvest of 1972 in Russia and the failure of the monsoon in India gave the meteorological world its first inkling of the global significance of El Niño. Historical records of the warming of the waters off the coast of Peru go back as far as 1525, but the Chimu people knew about the repercussions that this could have on fishing long before that. In 1972, the anchovy harvest, which reached 10 million tons in the previous year, fell to little more than 100,000 tons. It has never recovered, so destroying the largest fishing industry in the world. Now we know that what was at first regarded as a local catastrophe has global implications. When there are El Niño conditions in the eastern Pacific, there is usually drought in Australia, Brazil, India and southern Africa but floods in the Horn of Africa and along the western coast of South America, as well as irregular weather patterns at higher latitudes. El Niño has also contributed to disastrous fires in the rainforests of the Amazon and Borneo.

CYCLES AND LONG-TERM FORECASTS

UNDERSTANDING THE CYCLES THAT DOMINATE OUR
CLIMATE MAY HOLD THE KEY TO LONG-TERM WEATHER
FORECASTING OF THE FUTURE.

Climatologists have long been fascinated by regular cycles that seem to determine our climate. This is still a grey area, but identifying reliable cycles and their causes would enable meteorologists to make forecasts years ahead.

EVIDENCE OF CYCLES

Scientists have studied the huge mass of weather statistics looking for evidence of cycles. Once they have established regular cycles, they have to identify the possible causes – whether they are due to the natural variability of the climate or external factors, such as solar, lunar or planetary activity.

Apart from the daily and yearly cycles, the best example of a weather cycle is the Quasi-biennial Oscillation (QBO) in the stratosphere. This regular switch in the upper wind direction over the equator has been closely studied since the early 1950s and in recent years links have been found between this periodicity and weather at lower atmospheric levels. In particular, it has been linked with the fluctuations in El Niño and the frequency of hurricanes in the Atlantic.

Other much-discussed cycles are the sunspot fluctuatuions every 11 and 22 years. Meteorologists have found evidence of the latter in climate records: global marine-temperature time series, Greenland ice-core data, the Central England temperature record and the rainfall records for the prairies of North America.

Evidence has also been found for a cycle of 100 and 200 years that may be linked to solar activity. On the millennial timescale there is relatively little of note, except for the variation in the path and position of the Earth's orbit between 10,000–100,000 years, which seems to have been the main factor driving ice ages over the last million years or so.

WHAT CAUSES WEATHER CYCLES?

The problem with looking for cycles in weather statistics to correspond to the sunspot cycle is that variations and fluctuations may be caused by other factors, such as lunar tides or the natural interaction of the various components of the climate system. The oceans and atmosphere do interact to produce quasi-cyclic changes in the climate – changes that swing back and forth, or come and go like 'Will o' the wisp'. El Niño and La Niña are examples of this. It is because of the climate's erratic nature that it is so important for meteorologists to find clear links between any observed fluctuations and known physical causes. This is why the connections between the Earth's orbital variations and the ice ages are so convincing.

LONG-TERM FORECASTING DIFFICULTIES

Apart from the QBO and the ice ages, none of the identified cycles is sufficiently reliable to be used for forecasts. In order to predict what the weather will be like more than a year or so ahead, meteorologists need a better understanding of what causes the quasi-cyclic and chaotic variations in the climate system. The most convincing proposal suggests that, when a major disturbance disrupts the climate system, it can produce a quasi-cyclic response, which causes systems to swing to and fro before regaining their natural state. This theory is best considered in terms of ENSO in the Pacific, which swings back and forth every few years and will be tested following the 1997 development of El Niño.

SEARCHING FOR CYCLES

Spectral analysis of tree-ring series has enabled scientists to identify cycles of varying lengths. The longer the cycle the more variable it is.

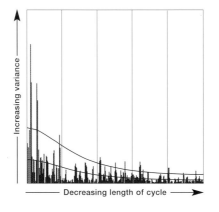

Increasing variance

Decreasing length of cycle →

THE QUASI-BIENNIAL OSCILLATION

There is a cycle, associated with the fluctuations in the position of the ITCZ, which goes on high above the seething region of convection in the tropics. The Quasi-biennial Oscillation (QBO) is the periodic reversal of the winds in the lower stratosphere at levels of around 20–30km (65,000–100,000ft) over equatorial regions. Every 27 months or so, these winds switch direction and turn from blowing strongly easterly to blowing nearly as strongly westerly. This reversal of the winds first takes place high in the stratosphere and propagates downwards over several months. This is shown below: fluctuations in wind speeds at altitudes of 20km/65,000ft (blue) follow a few months after those that occur at heights of 23km/75,000ft (red). The QBO is the only well-established weather cycle longer than the dominant annual cycle. It also appears to affect many features of the weather at lower levels and higher latitudes. For instance, when it is in the westerly phase it enhances hurricane activity in the Atlantic, whereas when it is in the reverse phase it dampens down activity. The cause and movement of the QBO is very complicated and can only be reproduced using special computer models that are more sensitive than usual modelling computers.

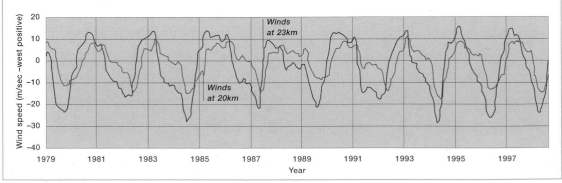

THE 1997 EL NIÑO EVENT

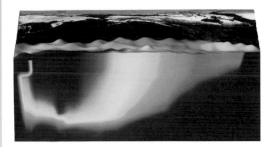

EARLY EL NIÑO (JANUARY 1997)
The temperature values range from 30°C (86°F), marked in red, to 8°C (46°F), in dark blue.

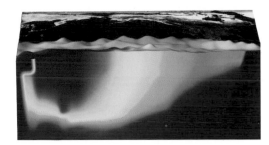

DEVELOPING EL NIÑO (JUNE 1997)
The water starts to spread across the western Pacific as a long warm tongue.

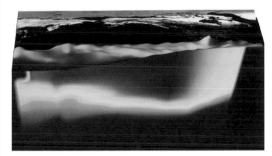

FULLY DEVELOPED EL NIÑO (NOVEMBER 1997)
During this El Niño a warm mass of water in the Pacific extends all the way to the coast of South America.

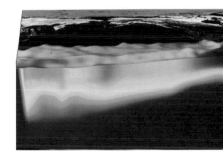

WANING EL NIÑO (MARCH 1998)
The temperatures start to drop across the Pacific as El Niño begins to wane.

SIR GILBERT WALKER (1868–1958)
Walker was Director General of the Indian Department of Meteorology from 1904–24. He identified three large-scale examples of 'see-saw' atmospheric patterns, which are valuable in long-range forecasting, especially in respect of the Indian monsoon. These were based on swings in pressure: the Southern Oscillation between the tropical Pacific and the Indian Ocean, the North Atlantic Oscillation between the Azores and Iceland, and the North Pacific Oscillation between Hawaii and southern Alaska. Although these patterns did not lead to forecasting successes at the time, they have re-emerged as important in understanding longer-term fluctuations in the climate.

Meteorologists have also carried out studies on non-linear systems – the chaos theory – and have found clear evidence that the climate is capable of switching between different states abruptly and frequently. These changes are by definition unpredictable: we can only hold our breath and hope that neither natural fluctuations nor human activities trigger any such dramatic shifts in the Earth's climate.

SEASONAL FORECASTING SUCCESSES

While forecasting events that will take place years ahead may still be in the realms of fantasy, seasonal behaviour up to 12 months ahead may be predictable, notably by finding connections between slowly varying components of the climate, such as sea surface temperatures and global weather patterns. Sir Gilbert Walker's (see box, left) early work on these connections has recently borne fruit in progress on predicting the behaviour of ENSO.

El Niño undoubtedly affects seasonal weather patterns around the whole globe, causing drought or abundant rainfall in places as far apart as Australia, Brazil, India and Zimbabwe and an increase in tropical cyclones, especially in the tropical Atlantic and perhaps even at higher latitudes. Therefore, predicting the future behaviour of El Niño is not limited to the equatorial Pacific but has become a matter of global importance. Forecasting is based on computer models and empirical methods, both of which have had some success. However, these wind patterns and the switches between them are notoriously difficult to forecast and so the prediction of El Niño events is still limited by the inherent uncertainty of modelling the behaviour of the atmosphere.

As we cannot always predict the chaotic behaviour of the atmosphere, it is often more worthwhile to concentrate on making better measurements of the less erratic oceans, using a vast network of buoys. These provide unequivocal evidence of the scale of sea surface warming and of the increasing depth of the thermocline in the eastern Pacific during the spring of 1997.

So, while the long-term prediction of El Niño events may be limited by the behaviour of the atmosphere, the inertia of the ocean is such that, once an event is clearly underway, its subsequent evolution is far easier to predict.

Left: The dead iguana is an example of how vulnerable wildlife on the Galapagos Islands in the equatorial Pacific is to the exaggerated swings in temperature and rainfall associated with the waxing and waning of the El Niño.

Modelling the climate 170–1

DESERT REGIONS

CLIMATE OF THE DESERTS

THE UNDULATING SAND DUNES AND HILLS OF THE DESERTS MAKE UP 10 PER CENT OF THE EARTH'S SURFACE AND HAVE A MAJOR EFFECT ON THE CLIMATE.

LUXOR, EGYPT

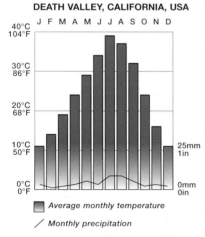

DEATH VALLEY, CALIFORNIA, USA

■ *Average monthly temperature*

╱ *Monthly precipitation*

Right: The Sossusvlei Dune in the Namib Desert, Namibia, is a distinctive feature of the desert landscape.

Desert regions play an important part of the global climate system because they are highly reflective and so can alter the radiative properties of the Earth. Their fierce climate is one of extremes and poses some of the greatest climatic challenges for people living there.

WHAT DO WE MEAN BY DESERTS?

We all know what deserts look like. They form the backdrop to *Lawrence of Arabia, Beau Geste* and *The English Patient* and consist of a sea of sand bathed in blistering sunshine. Alternatively, they are the rocky barren landscapes of American westerns dotted with imposing cacti, scrubby creosote bushes and tumbleweed. In practice, however, it is not easy to specify how little rainfall is required to constitute a desert. If we stick to the hot deserts of the world (for example, leaving out the colder wastes of central Asia and the Great Basin of western North America) then an annual figure of less than 300mm (12in) of rain is often used to draw the line. By this standard, deserts may include some relatively heavily vegetated regions where, for example, there is adequate groundwater from adjacent upland areas to support plant life. This measurement may also exclude some extremely hot, fairly arid places.

Perhaps the key feature of real deserts is the extreme unreliability of the rainfall. Any average statistics fail to reflect the unpredictable nature of the rainfall events. Often it rains very heavily for a few minutes or hours but then there may be no rainfall for days, months or even years. These heavy downpours sometimes coincide with the rainy season in adjacent semi-arid regions but it is impossible to pin them down to a precise time each year. In addition, rain is often highly localized, which means that some areas remain parched while nearby regions are flooded.

The only deserts that have a reliable source of water are those that lie alongside oceans with cold upwelling currents. Cold moist fog forms over these oceans and then rolls inland bringing moisture ashore, which is intercepted by plants and animals. The areas that benefit most from this process are the extremely arid coastal areas of Baja, California, the Atacama region of Peru and Chile, and the Namib Desert of south-western Africa, where the condensation can be equivalent to some 50–300mm (2–12in) of rainfall per year.

EXTREME TEMPERATURES

Because deserts are so hot by day, the temperature figures tend to attract a great deal of attention. But, in many respects the high temperatures do not represent the most striking feature of the hot deserts. True, the highest figures ever recorded have been in deserts – for example 57.8°C (136°F) at Al' Aziziyah, Libya, and 56.7°C (134°F) in Death Valley, California. But it is the combination of extremes that makes deserts climatically so interesting. Typical average highs in the hottest months are around 40–45°C (104–113°F), while the average lowest temperatures are around 20–25°C (68–77°F), but within 24 hours the temperature can vary by more than 30°C (54°F) and this change has been known to be as high as 50°C (90°F). This large range of temperatures only affects the ground to a depth of around 50cm (20in). Below this, the temperature tends to remain constant, rising no higher than the annual average figure of around 30°C (86°F). This is vital for the survival of desert plants and also animals, which can escape from the searing heat by retreating underground. Equally important is that, where there is almost no vegetation, the maximum temperature depends on the type of terrain and colour of the rocks and sand. It is estimated that in steep-sided rocky valleys in the hottest conditions the air temperature may exceed 65°C (149°F).

EVAPORATION RATES IN DESERTS

Not only is water an extremely rare commodity in the desert, but any water that there is evaporates very quickly. Evaporation rates depend on temperature, the relative

Above: This typical desert vegetation in the western USA is in the Mojave Desert Preserve, California.

Right: A Landsat false-colour image of sand dunes in Rub al Kali between Oman and South Yemen. The area is a closed drainage basin with long wind-blown sand dunes (yellow) moving over a mixture of clay, silt and muddy sand (blue).

humidity and wind speed. In most cases, potential water loss is high because it is dictated by the high daytime soil temperatures, which are often 10–20°C (18–36°F) higher than the air temperature. The drying effect of the heat is accentuated by the fact that the desert air is itself very dry and wind speeds are relatively strong. The rapid fall in temperatures at night can, however, lead to the formation of morning dew that some plants and animals use to survive.

Measurements of the rate of evaporation from water surfaces show that hot deserts lose more than 250cm (100in) per year, and in extreme cases as much as 400cm (160in) per year. Therefore any replenishment of soil moisture is rapidly lost to the atmosphere. However, if it penetrates to any significant depth the evaporative process will be slowed because a dry insulating surface layer quickly forms. The nature of the soil defines how much rain percolates deep down, with the right soil water here providing a life-giving source for many desert inhabitants.

In the driest places, such as the central Sahara, where average rainfall is less than 20mm (⅘in) per year, permanent vegetation is limited to isolated wadis (ephemeral river channels), which collect the limited rainfall. For the rest, the combination of aridity, intensity of evaporation, drifting sand and sometimes salinity means no vegetation can grow.

ALBEDO AND RADIATIVE EFFECTS

THE EARTH REFLECTS SUNLIGHT AND, AS THE MOST
REFLECTIVE SOIL SURFACE OF ALL, DESERTS ARE
ESPECIALLY IMPORTANT TO THE CLIMATE.

The hot deserts of the world are often regarded as a growing contributor to global warming. In practice the role of deserts in the climate is a more subtle process that needs to be looked at most carefully.

HOW DO DESERTS AFFECT THE CLIMATE?
The presence of extensive areas of open land has a significant impact on the local climate. Lack of vegetation increases the albedo of the ground: deserts reflect 30 per cent of the sunlight back out to space as opposed to the 14 per cent reflected by savannah grassland. This should, in theory, have a cooling effect on the climate. However, desert sand absorbs less heat and so generates less convection in the air, producing fewer clouds – effective reflectors of sunlight at the atmospheric level. This means that, considering the overall reflection from Earth and atmosphere, the net cooling effect is not as significant. At night, however, the lack of clouds means that the air temperature drops rapidly and a great deal of heat is lost to space. Indeed, measurements made by satellites show that outgoing radiation over the Sahara fractionally exceeds incoming solar radiation. Therefore, while deserts are hot places, they can potentially have a cooling effect on the climate, especially if their extent increases.

CAN THE DESERT SUSTAIN ITSELF?
The radiative effect of deserts was also invoked to explain the prolonged nature of the Sahel drought. One argument, popular in the 1970s, was that any change in the amount of sunlight reflected back to space would produce a runaway positive-feedback effect by reinforcing the process of desertification. Because it absorbs less heat, desert sand generates less convection in the air, producing fewer clouds and enfeebling the rainy season. Therefore, once desert is created it remains as desert. However, subsequent satellite observations did not support this theory because vegetation rapidly regenerates when the rains return, suggesting that large-scale weather fluctuations are more important than the runaway theory in maintaining desert conditions. This springing back to life occurs because seeds are capable of lying dormant for many years. They cannot, however, lie dormant forever and so long-term climate shifts may well be reinforced by the radiative effects of desert regions.

WHAT WERE DESERTS LIKE IN THE PAST?
High in the Ahaggar of the central Sahara are some beautiful prehistoric frescoes showing giraffe, elephant, cattle and herdsmen. These images are graphic evidence of how the aridity of the world's largest desert has changed since the last Ice age. Around 6000 years ago, at the peak of the Holocene warm spell, much of the Sahara received more rainfall than it does at present. Studies of the extent of Lake Chad show that, when it was at its highest level soon after the end of the Ice age, it had an area of

Right: The sparse vegetation in Death Valley, California, USA shows how reflective the desert's surface is.

Left: A boat marooned on the dried-up salt flats of what were once the shore of the Aral Sea, Kazakhstan. In places the shoreline of the sea has receded by more than 120km (75 miles).

400,000km² (150,000 square miles), and this declined very slowly until around 4000 years ago, when the climate of the region became noticeably drier. Computer models of the climate, which seek to simulate the conditions of this wetter period, show that the monsoon rains were much more extensive 6000 years ago. For the last four millennia, in spite of significant short-term fluctuations, there has been a steady increase in the desiccation of the region. By 1963 the area of Lake Chad was 23,500km² (9100 square miles) and by 1985 this figure had fallen to only 2000km² (770 square miles).

IRRIGATION AND SALINIZATION

Efforts to irrigate the soil in desert regions have caused serious problems with salinization – high levels of salt that most plants are not able to tolerate. This happens because water in the soil evaporates very quickly and any dissolved mineral salts are left behind. Salinization used to be the curse of many ancient riverine civilizations in North Africa, the Middle East and Asia. As early as the end of the third millennium BC, records in Mesopotamia tell of declining agricultural yields, moves to more salt-tolerant species such as barley and the widespread loss of land to cultivation in an attempt to increase harvests. Attempts over the last century or two to exploit arid areas for agricultural purposes have run into the same sort of problems.

The dangers of salinization are most profound where irrigation raises the water table close to the ground surface. Attempts to control salinization by dropping the water table and flushing the salts out can sometimes make matters worse, as this alters the balance of the different salts in a way which is even more damaging to plant life.

Salinization can also have a drastic impact on any river systems and seas that are nearby. The most dramatic example of this can be seen in the desiccation of the Aral Sea in Kazakhstan. Water was taken from its main feeders, the Amu Darya and Syr Darya rivers, to irrigate fields and this had disastrous consequences for the surrounding area. In the 30 years following the late 1950s, the surface area

Below: An ancient rock engraving of people and cattle at Jabbaren, southern Algeria. Jabbaren is located in the Tassili range of the Ahaggar in the heart of the Sahara.

of the sea more than halved to 35,000km² (13,500 square miles), its volume fell by more than two-thirds to 300km³ (72 cubic miles), and its salinity tripled. The exposed seabed has become a salty wasteland, as have several million hectares of cotton fields that were previously irrigated by the two river systems.

RENEWABLE ENERGY

RECYCLING OF PRODUCTS SUCH AS BOTTLES AND TIN CANS IS NOW COMMON, BUT HOW CAN WE RECYCLE THE MOST FUNDAMENTAL RESOURCE – ENERGY?

The abundant supply of sunshine in desert areas makes them reliable sites for exploiting the potential of solar power. The methods that are used to harness the Sun's rays also reveal much about the possibilities of utilizing other forms of recycled energy.

We already use a great deal of renewable energy. Currently 15–20 per cent of total world energy demand is met from these sources. Hydro power meets 20 per cent of the world's electricity requirements: it supplies around 50 per cent of the needs of Sweden and New Zealand, 60 per cent in Canada and nearly 100 per cent in Norway. In developing countries in Africa and Asia, biomass (organic matter) is a particularly important resource, supplying over 90 per cent of total energy in Nepal and Malawi and between 25 and 50 per cent of the needs of large industrial countries such as China, India and Brazil.

AN ABUNDANT ENERGY SUPPLY

A fundamental question is how much renewable energy is available. The simple answer is that there is far more than we could ever need. For example, photovoltaic or solar-power systems, which convert sunlight into electricity could provide enough energy to power the whole of the USA. If these systems were at a highly effective rate of 15 per cent efficiency, they would only need to cover 1 per cent of the country's surface in order to produce enough energy for the whole country – the same area as is currently covered by highways and roads. For developing countries, covering less than 0.1 per cent of their land area could supply all their energy needs.

Therefore, there is no shortage of renewable energy, whether it is obtained directly from incoming solar radiation or by exploiting natural sources that are driven by the Sun, such as wind, waves, biomass and hydroelectric power. The fundamental issue is whether they can be exploited in a manner that is both economical and environmentally acceptable. This can only be done if the manufacture and maintenance of systems are improved so that they are able to compete in the marketplace on equal terms with conventional energy sources.

SOLAR POWER

Solar energy systems take advantage of the least cloudy places in the world where there is abundant sunshine throughout the year. This is why much of the work on the potential of solar energy has been conducted in the desert of south-west USA. But complex technology and desert sunshine are not the only answers: simple technological equipment that can meet seasonal demand is already being widely used. For example, the tourist requirement for hot water in the evening in Mediterranean resorts is easily met by simple solar-heating systems, which exploit the midday sun. Similarly, in temperate continental regions, house design can take advantage of the low-angle sunlight in chilly mid-winter and provide shade from high-angle rays in steamy summer. For an aesthetically pleasing and economically worthwhile solution, this can be combined with the strategic planting of shade-providing deciduous trees that also let in winter sun.

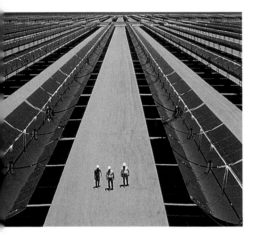

Above: One of three solar-energy complexes in the Mojave Desert, California, USA. Covering 400 hectares (1000 acres) in total, they can generate 275 megawatts of electricity (90 per cent of global solar production). This complex has 650,000 parabolic mirrors – these track the Sun across the sky and focus the light on to tubes of synthetic oil, which are heated to 391°C (736°F). This superheated oil is used to boil water and drive steam turbines, which generate electricity.

Right: Cleaning a photovoltaic solar panel in Waat, Sudan.

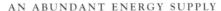

Left: A wind farm in Palm Springs, California, USA. The electrical energy generated from a wind turbine is proportional to the cube of the wind speed. Thus, a 40kph (24mph) wind will generate eight times more energy than a 20kph (12mph) wind. So the best place for wind turbines is in areas where there are frequent strong winds.

The economics of photovoltaic systems are still a long way from being able to compete with conventional power generation. But, where the level of demand is low and the electricity-supply systems are less widespread, the advantages of photovoltaic systems to provide lighting and refrigeration are already well-established. So the expansion of solar-power use is likely to build on these specific niche areas and it may eventually be able to compete with large-scale generator systems.

OTHER SOURCES: WATER, WIND, WAVES

The potential for exploiting water, wind and waves varies considerably and has a lengthy history. In medieval England there were some 6000 watermills, which provided power for grinding grain; while flat, breezy Holland is renowned for its windmills, which were used to drain its polders for centuries. Recent years have seen great advances in wind farms (see above), which are now used worldwide.

Hydroelectric schemes have been built throughout the world to exploit the power of major river systems though there are important environmental considerations surrounding this energy use, such as flooding of large areas of arable land and retention of valuable silt. In addition, there are doubts as to whether there are many new rivers that can still be exploited around the world. However, hydroelectric power can provide substantial additional energy in many countries, especially in the developing world.

The economics of windmills depend to a large extent on the frequency of sufficiently strong winds. For this reason, places that lie in the track of mid-latitude westerlies are the obvious locations for development of modern windmill systems – they are particularly popular in Britain and Denmark. Prevailing winds are also important in producing waves of sufficient size to power generators, but economically viable wave-power systems have yet to be tried and tested.

BIOMASS: ORGANIC FUEL

As plants grow, they absorb carbon dioxide to form their stems, branches and leaves, and store carbon in the ground in their roots. Therefore, the use of plants as a fuel has the additional benefit of storing carbon in the ground. Biomass, which includes fast-growing trees such as the willow tree, can be harvested directly to provide fuel. Alternatively, this organic matter can be collected as the by-products of forestry, agriculture and animal husbandry or as part of domestic and industrial waste, for use either directly as fuel or for anaerobic digestion to produce methane, which is a much cleaner and more flexible fuel.

PREVENTING GLOBAL WARMING

Many of the arguments for developing renewable energy are bolstered by the extent to which they will avoid the build-up of greenhouse gases in the atmosphere, as happens with fossil fuel combustion. Forecasts of the future contribution of renewables to global energy supplies vary substantially, depending on the growth rate of energy demand and on the economics of different forms of supply – one estimate suggests that as much as 45 per cent of our energy could come from renewable sources by 2050.

The next 100 years 184–5

DESERT FLORA AND FAUNA

THE PLANTS AND ANIMALS THAT LIVE IN THE HOT DESERTS OF THE WORLD HAVE EVOLVED AN AMAZING RANGE OF FINELY TUNED COPING STRATEGIES.

Desert herbaceous plants can survive temperatures of up to 50–55ºC (122–131ºF), while the prickly pear can stand a scorching 65ºC (149ºF). Animals have a lower tolerance level and most species find themselves in real difficulty when the temperature rises to approximately 40–50ºC (104–122ºF), which is why so many of them spend the daytime buried underneath the ground. But some beetles and scorpions can comfortably survive conditions up to 50ºC (122ºF).

AVOIDING THE HEAT

There are two principal strategies in heat survival: avoid the heat or tolerate it. The most common method for both plants and animals is to hide away from the heat of the day.

off germination. Rapid growth then follows, producing sufficient seeds for the next generation quickly while supplies of water last.

TOLERATING THE HEAT

Even in the most arid of conditions there are tolerant forms of fauna and flora that are able to survive extreme heat by controlling their temperature and water loss. Animals adopt several strategies in order to survive the extreme heat: they expose as little of their surface area as possible to the sunlight; they have often evolved to be light in colour, which minimizes absorption and maximizes reflection of solar radiation; and they often have a surface growth of spines or hairs, which absorbs or reflects much of the incident radiation and creates a boundary layer to insulate the underlying surface. Large animals, such as the camel and goat, rely on evaporative cooling. The camel also has broadened feet for walking on sand, thick eyelashes and nostrils that close to keep sand out and thickened lips to

Below: Dromedary (one-humped) camels in the Australian outback. They were first imported from India and Afghanistan in the 1840s.

Many plants live underneath the shelter of shrubs, in moist cool nooks and crannies, or below ground. Alternatively, they adopt lifecycles that enable them to survive long periods of dormancy, ready to burst into activity when more clement conditions arrive. Some plants have seeds that can lie dormant for many years until the right conditions (for example, abrasion caused by flash floods) occur to trigger

enable it to eat the coarsest prickliest vegetation. The date palm lizard of the Sahara changes colour according to the temperature, from blackish-brown in cooler conditions to yellow-orange or green in hot weather to reduce absorption. One particularly well-adapted variety of spider builds webs in the Namib Desert near the Atlantic coast to collect condensation from coastal fogs. Jerboas (desert rodents)

153 Evaporation rates in deserts

have extremely long back legs to enable them to travel efficiently over sand in rapid bounds in search of food. The horned viper has special scales down its side which enable it to shimmy down into the sand and lie in wait for its prey. Some birds get away from it all by soaring on thermal wind currents to where the air is noticeably cooler.

SURVIVING WITHOUT WATER

Animals use a number of methods to preserve water: many do not perspire very much, their excreta tends to be dry and they only produce small amounts of very concentrated urine. Some can tolerate substantial dehydration over lengthy periods. The camel, for example, when deprived of water, can tolerate a 25 per cent loss in body weight. In high summer this loss takes seven days, whereas in winter going without water for 17 days only reduces its weight by a mere 16 per cent. When camels get the opportunity, they have the extraordinary ability to drink large quantities of water in minutes to replenish lost reserves. Similarly, a few birds, some plants and many insects can survive a high degree of tissue desiccation.

Plants also have a wide range of strategies to reduce water loss. Their stomata only open at night or are sunken, and they have thick cuticles and surface hairs. A low surface area to volume ratio helps, which is why many cacti are roughly spherical in shape. Some plants, such as the creosote and many cacti, have a large volume of shallow, lateral roots, which exploit brief rainstorms efficiently; and others, such as the mesquite, have roots that penetrate 10–30m (33–100ft) down into the substratum where they can tap permanent sources of groundwater. Extensive root systems are the method by which individual plants compete with each other to obtain adequate water, and explain why many mature desert shrubs tend to be rather uniformly distributed across the landscape.

Above: The desert cactus growing in Peru.

Left: The thorny devil, which is able to survive in the extreme heat of the desert of central Australia.

The key to survival: water 161

HUMAN RESPONSE TO HEAT

HUMAN BEINGS ARE FAR BETTER EQUIPPED TO
HANDLE THE CHALLENGES POSED BY EXTREME HEAT
THAN THOSE PRESENTED BY EXTREME COLD.

The first human beings on the Earth lived in the hot, dry conditions of the savannahs of Africa; and evolved to survive in such conditions.

SWEAT – THE HUMAN THERMOSTAT

Acclimatizing to the heat is a universal human capability. We keep cool by sweating – as our perspiration evaporates, it uses body heat in the evaporation process and so, as a result, our skin temperature lowers. The surface of our skin is covered by more than 1.5 million sweat glands that can produce copious quantities of perspiration over the whole body. Our relative lack of body hair speeds up evaporation; the price we pay is the risk of dehydration.

DEHYDRATION

Activity leads to swifter water loss. Tests carried out in the Arizona desert show that a man walking briskly in the sun loses around 1 litre (1.7 pints) of sweat per hour at 35°C (95°F) and the figure rises to nearly 1.5 litres (2.5 pints) at 45°C (113°F). When sitting nude in the sun a man loses only 500ml (0.8 pint) per hour at 35°C and 900ml (1.5 pints) at 45°C, and these figures fall by about a further 50 per cent if he is sitting in the shade. Women sweat less than men in severe heat. Without regular supplies of water, we can rapidly dehydrate. Losses of up to 10 per cent of body weight cause severe thirst and eventually mental derangement but can be rapidly reversed by drinking water. Higher losses require medical help and the lethal limit of dehydration is probably 15–25 per cent of body weight.

ADAPTATION AND ACCLIMATIZATION

The indigenous people of the hottest, driest parts of the world have evolved physically in a variety of ways that seem related to the climate. Low body mass, long limbs and slim bodies make the cooling sweating process more efficient. Long noses may help to humidify the air and reduce water loss from the lungs, and pigmented skin provides some protection from the sunlight. This is adaptation – people whose bodies have adapted to the climate over generations.

However, medical tests show that all of us can acclimatize to the heat relatively quickly. People from temperate climates who travel to hotter parts of the world may feel very uncomfortable at first but, within 7–10 days, the blood volume increases, the sweat rate rises and becomes less salty, the heart rate drops (putting the cardiovascular system under less stress) and the body temperature does not rise as high with exercise. Thereafter the acclimatization process may go on for several years, by which time people's body systems will behave in the same way as those who have lived all their lives in hot places. So, when visiting a hotter part of the world, take it easy for the first week or so, after which you should be on a par with the locals. Women seem to be affected more than men by short-term exposure to heat and take a little longer to acclimatize fully.

Right: A Tuareg in the Sahara desert, Algeria. His loose-fitting clothing helps him remain comfortable in the blistering heat of the day.

Under extremely hot desert conditions the best way to acclimatize is to wear loose-fitting clothes that cover the arms and legs and prevent the absorption of solar radiation. Ideally, these should be white to reflect as much of the sunlight as possible but, because white materials are often translucent and allow heat – including ultraviolet rays that are responsible for sunburn – to penetrate, dark fabrics tend to be better. The voluminous, loosely worn thick wool or mohair robes worn by desert peoples are admirably suited to the purpose. They act as insulation against heat gain during the day, provide adequate ventilation to enable evaporative cooling by sweating to occur and insulate against heat loss at night.

THE KEY TO SURVIVAL: WATER

Unlike some animals, humans need regular supplies of significant quantities of water in order to be able to survive and so the challenge of living permanently in the most arid parts of the world is all about one thing – water. In some regions, the scarcity of water and the sparseness of vegetation mean that native people have had to live a nomadic existence, moving from place to place in the search of food and water for themselves and their animals. Where there is an adequate supply of water, at oases, a fertile island can be created in the desert and living conditions for people will be good.

DESERT DWELLINGS

In the hottest parts of industrialized countries, air-conditioning is widely used to control the baking heat of summer. However, simple solutions, such as clothing and building design, can provide relief from all but the worst extremes. In buildings, the use of thick adobe or brick walls provides insulation from intense heat, providing a constant temperature instead of the wide variations outside, and taking a leaf out of nature's book by imitating the many species that spend the baking daylight hours underground. In the desert areas of the south-west USA, for example, buildings of this design are pleasantly cool by day and relatively warm at night.

In parts of the Middle East and the Indian subcontinent, buildings are built in positions that take advantage of wind or sea breezes: wind catchers attached at right angles to chimneys deflect the gusts and produce cooling draughts inside. Indeed, according to papyrus records, wind catchers were used on houses in Egypt as long ago as 1500BC. Alternatively, chimneys are designed so that, when they heat up in the midday sun, the warm air rises and air is pulled up from the room below, creating extra draughts even during calm conditions. In deserts, such as Rajasthan, India, inhabitants hang reed mats in the doorway and moisten them during the heat of the day. Any draught is then cooled by evaporation – natural air-conditioning.

THE SAHEL DROUGHT

The images from the Sahel in the early 1970s, when over 100,000 people died, or from the Ethiopian famine in 1984, which had an even greater mortality rate, brought home to many people in the developed world the human consequences of climate change.

The drought in the Sahel began in 1968 and reached its first peak in 1972. It abated to a certain extent but returned in the late 1970s and again with greater vigour during the 1980s before easing off during the 1990s. In spite of the fact that drought is a natural feature of the region, the events in the 1970s and thereafter have had a major impact on many people's perceptions about the climate. Drought in sub-Saharan Africa has come to represent both the threat of global warming and the demands that growing populations inevitably place on limited natural resources.

What causes drought?

Droughts are most common in Africa but have to be considered in the context of global weather patterns. The large amounts of heat that are received by tropical regions generate convection – air rises over the equator and sinks at about 10–20°N and S. This sinking air is extremely dry and maintains the desert regions that exist in these bands of latitude. The amount and frequency of rainfall along the southern fringes of the Sahara depend on how far north the region of rainy convective activity – the intertropical convergence zone (ITCZ) – spreads during the summer months.

The climate of the Sahel is defined by this annual motion. This semi-arid region stretches in a narrow band across Africa from Senegal and Mauritania to the Red Sea. Rain only falls from June to September and amounts vary from 200 to 800mm (8 to 32in) per year from the north to the south of the region. These figures vary noticeably from year to year according to how far north the ITCZ spreads. The movement of the ITCZ is influenced by a variety of factors, most notably tropical sea surface temperatures. Therefore, Sahel rainfall is by nature erratic and subject to remote forces that we do not yet fully understand.

Predicting drought

Meteorologists have been faced with the formidable challenge of explaining why droughts happen and what part the fluctuating climate in the tropics and El Niño play in the process. They have used links that they have found between the drought in the Sahel desert and various meteorological factors around the world to put together seasonal forecasts for the sub-Saharan region with a considerable degree of success. It is possible to tell how dry the Sahel region will be in any year but they

Left: A local inhabitant digging a well as part of an attempt to supply water for the Kalsaka village in Yatenga Province, Burkina Faso.

cannot predict precisely where or when the rain will fall. The research does show, however, that accurate forecasts of detailed rain patterns will be possible at some stage in the future. Another approach has been to explore how global computer models handle the rainfall across the Sahel region when linked to sea surface temperatures. This method has the potential to produce good forecasts of seasonal rainfall, providing it takes the June figures into account – but this is a significant practical shortcoming because forecasts that are issued in July come far too late to influence planting decisions for the year.

Preparing for drought

Accurate and early drought forecasts are important both in terms of planning agriculture and delivering aid to the Sahel region. The economic benefit depends on who uses the information, ranging from international aid organizations, through the various local governments to the farmers, pastoralists and freshwater fishermen who are most affected by the rainfall fluctuations. Farmers clearly need forecasts in good time to decide which crops to sow. Therefore, rainfall forecasts need to be reduced to a form that can be broadcast by radio, reflects local calendars, fits in with other factors influencing agriculture and explains the potential strengths and weaknesses of predictions. Current forecasts fall well short of these demands, and in the foreseeable future it is likely that only large commercial growers with irrigated land will benefit from the predictions. At the other end of the scale the poorest producers, who lack the necessary funds to respond to forecasts, are likely to become more vulnerable to how the markets respond to the information and will therefore be in greater danger of impoverishment.

The benefits of forecasting an exceptionally dry season show up not only in planting decisions but also in livestock management. A national plan of herd destocking would completely alter the lives of the pastoralists, enabling them to sell off their livestock in a staggered manner. This would prevent huge surpluses from coming on to the market and would therefore maintain steady prices for the livestock. For donor agencies and governments the potential to exploit the forecasts is greater. Reliable forecasts are needed in good time because donor agencies have to allocate their aid resources early in each financial year. Aid and speedy responses from the local governments are most important in extreme years, and so accurate early predictions of very dry or very wet years are vital.

The massive aid-agency response to successive droughts in Africa has been a measure of public concern in the developed world, but has also raised sensitive issues of how best to provide help. The particular challenge in Africa is to reflect the state of economic development of the countries suffering from drought, as well as recognizing the demanding nature of the climate. Drawing on the much lengthier experience of providing aid in India, it is essential that aid agencies are equipped to make early interventions, drawing on local knowledge, to prevent matters getting out of hand. Wherever possible, aid should provide employment that generates income, and loans for farmers and pastoralists to buy seeds and livestock. The alternative of a drift into a dependency culture and accelerated migration to urban areas is much more damaging in the long run. These longer-term aims must fully involve the government of the country, but, even if successful, will take many years to bring to fruition and will always be vulnerable to climatic fluctuations.

ARE THE DESERTS EXPANDING?

THE SEEMINGLY INEXORABLE ADVANCE OF THE SAHARA HAS LED TO DEBATE ON DESERTIFICATION AND THE EXTENT TO WHICH IT IS DUE TO HUMAN ACTIVITIES.

The expansion of the deserts has made people think that a permanent shift in the climate is underway. However, it is all too easy to attribute excessive significance to what look like permanent changes, but which may prove to be nothing more than normal fluctuations when viewed over longer periods of time.

THE EVER-ADVANCING SAHEL

Desertification is not a new phenomenon. However, the changes that occurred in the Sahel in the 1970s seemed to be particularly extreme. Terrifying figures became accepted environmental statistics: namely that more than some 20 million hectares (50 million acres) of once-productive soil were being reduced to unproductive desert each year. This is equivalent to well over half the area of the British Isles or the whole of Kansas. The image of the Sahara marching southwards at up to 50km (30 miles) per year galvanized many aid agencies into action. This global concern culminated in the UN Conference on Desertification, held in Nairobi in 1977 (see below).

WHAT CAUSES DESERTIFICATION?

As we have already seen, the amount of rainfall in desert regions is determined by the movement of the intertropical convergence zone (ITCZ). The scale of the changes in desert extent in the 1970s led to speculation that there might be a permanent shift in the position of the ITCZ. Several theories have been put forward by scientists as to what might cause such a shift.

Firstly, changes on a local level were thought to be responsible, particularly overgrazing by herds belonging to nomads. When vegetation was stripped away by cattle, there would be an increase in the albedo because desert sand reflects much more sunlight than savannah plants. This, in turn, would produce a runaway positive-feedback effect, which would reinforce the process of desertification because the desert would absorb less heat, which would reduce convection and prevent the ITCZ from moving as far north as it normally does during the rainy season (June to September). This would mean that once the desert was created it would remain as desert.

Data collected during the 1980s did not substantiate this explanation and suggested instead that the dominant role of natural fluctuations in the global climate had been underestimated. Many of the shifts in the desert were in fact largely due to annual fluctuations in rainfall. Satellite observations of the extent of vegetation during the rainy season was closely linked to these changes in rainfall – plant life soon grew back in wetter years. In the period 1980–95, satellite observations of the western Sahel showed that the desert expanded and contracted by about 300km (180miles) in response to changes in rainfall from year to year. Detailed studies at ground level also showed how areas that looked like complete deserts had the astonishing capacity to spring back to life when heavy rain fell and long-dormant seeds germinated.

The answer as to why the drought in the Sahel has gone on for so long will probably become clearer when we have a better understanding of global climate change. Sea surface temperatures and El Niño are bound to be included because of their role in defining tropical climate patterns. However, fluctuations in El Niño are far more rapid than are the changes in the desert, which take place over decades. This implies that the answer lies not in one single factor but in the combined effect of sea surface temperature changes on a far wider scale.

PREVENTING THE DESERTS' EXPANSION

The UN Conference on Desertification, held in Nairobi in 1977, launched a plan of action that funded projects to combat desertification, such as irrigation and planting of trees in endangered areas. These cost some $6 billion and did little to prevent desertification over the next 15 years, raising doubts as to whether the projects were aiming at the right target. There is a fear that the whole concept of

Right: An anti-erosion project in Burkina Faso, designed to help to prevent the spread of the desert.

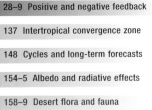

desertification was misconceived and what was needed were better measures of the changes that were actually occurring. In particular, there was no adequate distinction between degradation caused by human activities – such as overgrazing by pastoralists' herds, collection of firewood and inappropriate farming – and degradation caused by drought. If, as appears to be the case, the main factor is natural fluctuations in the climate, our most important aim is to develop strategies that are effective in responding to them. This does not mean that reducing the impact of human activities will not form part of the strategy, but that we must be realistic about what we can do when natural changes pose such a major challenge to any planning.

GREATER FORCES MAY BE AT WORK

We should learn from the experience of desertification not to overestimate one factor when analysing global change as a whole. Human activities must always be considered in conjunction with other factors. Sometimes there is one dominant cause for change and so it is wise to take all reasonable precautions to prevent doing lasting harm. But, in deciding how to use scarce resources, we must be on our guard against underestimating the capacity of many forms of life to adapt to sudden, massive changes in the climate. We should not forget that we have evolved through periods of far greater climatic variability than those of recorded history and that we possess genetic defences that enable us to survive a wide variety of extremes.

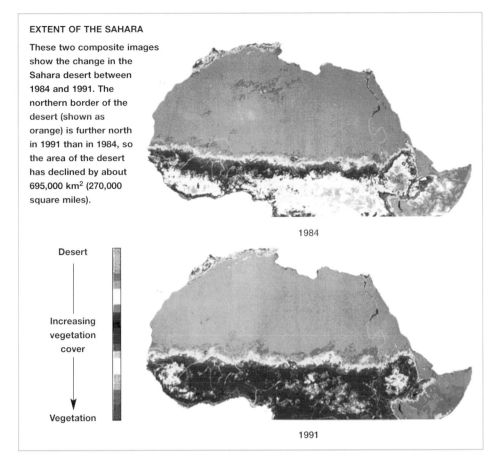

EXTENT OF THE SAHARA

These two composite images show the change in the Sahara desert between 1984 and 1991. The northern border of the desert (shown as orange) is further north in 1991 than in 1984, so the area of the desert has declined by about 695,000 km² (270,000 square miles).

Desert

↓

Increasing vegetation cover

↓

Vegetation

1984

1991

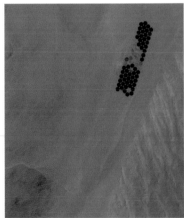

Above: The desert sands of Libya, as seen from the space shuttle Atlantis in 1990. The dark circles indicate areas of the desert that are being cultivated using centre-pivot irrigation systems.

Left: Sand dunes in the Namib Desert, showing how these waves of sand can threaten to engulf vegetation in adjacent areas.

How does dust affect the climate? 166

DUST AND AIRBORNE PARTICULATES

GREAT CLOUDS OF DUST, BLOWN FROM THE DESERTS OR THE BY-PRODUCTS OF INDUSTRY, MIGHT HAVE A COOLING EFFECT ON THE CLIMATE.

The effect that particles of sand whipped by winds from deserts, and particulates emitted by industrial processes have on the climate is the subject of continued debate.

HOW ARE DUST CLOUDS FORMED?

Dust clouds are formed in two ways. Firstly, even the most basic of farming techniques unsettles the soil and makes it easy for winds to blow it up into the atmosphere. The Dust Bowl years are an example of how inappropriate agricultural practices can cause dust clouds, with disastrous consequences for the local region.

Secondly, dust clouds can be generated when strong winds blow over desert regions. Saharan dust, for example, is often transported out across the tropical Atlantic, up into the Mediterranean and as far north as Britain. A recent study of satellite images taken between 1983 and 94 showed that the amount of dust was closely linked to the North Atlantic Oscillation. This weather pattern is linked to the strength of the Azores high-pressure system, which generates winds over North Africa.

Therefore, it is not yet possible to show whether the creation of dust in the Sahara is caused by human actions or whether it is due to the natural variability of the climate.

HOW DOES DUST AFFECT THE CLIMATE?

The many duststorms taking place around the world are having a substantial impact on the properties of the atmosphere. The impact is greatest where dust from arid regions streams out over the oceans. The dust clouds have a much higher reflectivity than the underlying ocean and therefore will have a cooling effect on the local climate. It is estimated that this cooling could be sufficient to cancel out about half the global warming that has taken place during the 20th century as a result of the build-up of greenhouse gases in the atmosphere. The areas where this cooling is likely to be greatest are over the tropical Atlantic to the west of Africa, and east of China.

These dust clouds may also have an interesting indirect impact on the climate in that they contain nutrients (nitrogen and iron), which fertilize otherwise unproductive areas of the ocean. This could stimulate the growth of plankton, which would mop up more carbon dioxide from the atmosphere and increase the number of particulates. This negative feedback could be a good example of the Gaia principle, which states that the biosphere behaves in a manner optimal for the maintenance of living things.

SULPHATE PARTICULATES

One of the most intensively researched aspects of particulate formation as a result of human activities is the creation of sulphate particulates by the combustion of fossil fuels. When these contain sulphur compounds, their combustion

Right: A view from space of a dust storm sweeping south across the Algeria–Niger border. These dust clouds can sometimes be transported across the tropical Atlantic as far as Barbados.

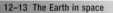

CONTRAILS

The clouds formed by high-flying aircraft, contrails, are one of the most visible aspects of human activities having an impact on the climate. These trails of ice crystals stretching across the skies can be seen both from the ground on clear sunny days and on satellite images. They can last for hours and even merge into each other to create a general thin haze. Unlike volcanic eruptions, which also inject large amounts of dust and aerosols into the stratosphere, the impact of aircraft is likely to have a warming effect. This is because the ice crystals formed by aircraft exhausts are much larger than the fine particles formed by volcanoes. This means that they are not very efficient reflectors or absorbers of sunlight coming in, but have a proportionately bigger effect on outgoing terrestrial radiation. In other words, they behave more like a blanket than a reflector. There are, however, few measurements that can provide evidence of this impact and so, as with many other aspects of human-generated dust and particulates, we need to do more research before there can be any certainty about the real impact of contrails.

produces sulphur dioxide, which is mainly converted into sulphuric acid in the atmosphere. This then combines with other compounds to form sulphates, which are efficient reflectors of sunlight. It is estimated that the amount of these tiny particulates that have been created as a result of the activities of the industrial nations of the world has had a noticeable cooling effect on our climate. This is the opposite effect of the burning of carbon in fossil fuels to form carbon dioxide, a greenhouse gas.

While the direct impact of sulphate particulates on the climate has been the subject of computer modelling work, any indirect effect these particles may have on the

Above: The smoke and particulates from industrial processes, such as this coal-fired power station in Brandenburg, Germany, are often referred to as 'the human volcano' because of their potential cooling effect on the climate.

formation of clouds has yet to be included in the models. The indirect impact of sulphate particulates on the climate is that they may lead to the more efficient production of clouds. This could increase cloudiness and hence bring about an additional cooling of the climate. We need a better measure of how much impact human activities are having on the nature and extent of clouds before we can fully understand this important phenomenon.

Global warming 184–5

THE FUTURE

MODELLING THE CLIMATE

WEATHER FORECASTS FOR THE WEEK AHEAD ARE MORE
ACCURATE THAN EVER, BUT HOW CAN WE PREDICT
WEATHER FOR NEXT YEAR OR EVEN NEXT CENTURY?

EDWARD LORENZ (1917–)
Lorenz, American mathematician
and meteorologist, is best known
for coining the expression 'the
Butterfly Effect', which has
become the icon of Chaos theory.
This reflects the fact that, in trying
to calculate the behaviour of the
atmosphere, tiniest uncertainties
in the initial conditions will
multiply in the computations to
swamp any predictions of their
future state. It is not simply that
a butterfly in Brazil may cause a
tornado in Texas, but the fact that
our computer models cannot
include what butterflies, or for
that matter jumbo jets, are doing
and so it will be impossible for our
weather forecasts ever to predict
more than about 10 days ahead.

The only way to predict how human activities may affect the climate is to develop computer models of the global weather system that reflect the current weather situation and can be used to simulate its behaviour for decades and centuries into the future. All our forecasts of global warming are based on this methodology.

GLOBAL CIRCULATION MODELS

Computer studies of climate change use what are known as Global Circulation Models (GCMs). These are similar in terms of mathematics and physics to those involved in the day-to-day weather-forecasting work. When considering large-scale shifts in the climate, GCMs have to simulate the behaviour of the oceans and how they interact with the atmosphere over longer periods of time, but with the same degree of accuracy as short-term weather forecasts. The supercomputers used are very expensive to run. To counteract this, meteorologists simplify the representation of the climate, cutting down on the detail by using a larger grid spacing on the model and by lengthening the time intervals between successive computations.

Incorporating an accurate representation of the oceans into these models is particularly demanding. The major currents and eddies are much smaller than corresponding atmospheric weather systems and so models of the oceans must have a much higher resolution than those of the atmosphere. The highest-resolution research models of the oceans have up to 60 levels in depth and resolutions as fine as one-sixth of a degree of latitude and longitude. While these can produce realistic representations of many of the oceans' dynamics, they are far too expensive and slow to incorporate into current climate studies.

Coupled atmosphere–ocean GCMs have a resolution of a few degrees of latitude and longitude for both components. This compromise poses problems for how the oceans and atmosphere are represented as exchanging heat, energy and water vapour in regions where there are significant variations in sea surface temperatures over short distances. Until recently, this lack of detail made models stray off track. The solution was for meteorologists to make the necessary adjustments to nudge them back on course. There are clearly serious scientific objections to this fudging of figures and so it is encouraging that the latest models are starting to show signs of behaving more accurately without any of these adjustments.

HOW ACCURATE ARE THE MODELS?

To date, 16 GCMs have been developed by the leading research groups worldwide and they have adopted many ingenious ways of addressing the complexity of the climate. This multiplicity of approaches provides an insight into which climatic processes make a big difference to how the models perform and where the models have most difficulty simulating the real world. Their progress has been reviewed by the UN Intergovernmental Panel on Climate Change.

The first challenge is to provide an overall view of what the future climate will be like, by predicting how temperature and precipitation will change. A simple way of testing how accurate the models are is to compare the figures that they predicted several years ago with actual observed values today. Their combined average-temperature forecasts have been proven to be relatively accurate, reflecting the recent warming trend. However, the results of the 16 models differ by as much as 6°C (11°F) – a significant margin in terms of both past and future climate change. With regard to precipitation, it is more difficult to check how accurately the models are reproducing the current state of affairs because we do not know exactly how much rainfall is occurring around the world at any one time.

A HOTTER WETTER WORLD

Overall, the GCMs predict that our world is becoming warmer and wetter – temperatures are expected to rise by 1.5–4.5°C (2.7–8.1°F) over the next 100 years and rainfall will increase, particularly in the tropics and mid-latitudes. Again, there are variations between the 16 models with regard to how intense the rainfall is, particularly in the tropics. These discrepancies are important because the precipitation rate affects how much water is passing through the hydrological cycle and the path of ocean thermohaline circulation (if there is enough rainfall, the salinity of the oceans may be changed sufficiently to alter the path of deep-water currents). Generally, the models that predict higher temperatures also predict most precipitation because, when it is warmer, more water evaporates and additional clouds form, resulting in an increase in rainfall.

SNOW, SEA ICE, OCEANS AND CLOUDS

Models also predict the behaviour of specific components of the Earth's climate system, including snow and sea-ice cover, ocean currents and cloud cover. These cause considerable difficulties for scientists because past predictions do not tally with observed changes. In terms of future snow and sea-ice cover, the results are clearly misleading: some models predict a great deal of winter snow cover, which lasts well into summer, while others lose all their sea ice in the Arctic and almost all their sea ice in the southern hemisphere throughout the year. So modellers have had to make adjustments in order to produce sensible conditions that are closer to our real experience.

Similar problems crop up in the handling of ocean currents because the various models forecast very different scenarios. Some GCMs predict that 10 times more warm water will be transported polewards in the North Atlantic by thermohaline circulation than others. The most accurate models are those that have had flux adjustment.

The ability of GCMs to handle clouds is also critical. The main issue is whether variations in cloudiness will reinforce any climate change (positive feedback) or counteract the change (negative feedback). Clouds exert a strong influence on the Earth's climate: they can stop heat reaching the Earth by radiating it or reflecting it back out to space, but they can also act as blankets, preventing terrestrial radiation from escaping. As such, the treatment of cloud type and distribution in GCMs has a major impact on whether the feedback is positive or negative. Early models predicted a strong positive feedback, which meant that global warming would be further enhanced by increased tropical cloudiness; more recent models have produced less extreme results.

There are still problems in GCMs' handling of clouds. We need a better understanding of the specific properties of different types of clouds in order to be able to simulate their behaviour accurately. Similarly, it is difficult to represent global cloud cover in GCMs, particularly in terms of exactly where clouds form and at what time of year.

There is yet one more problem with modelling clouds, which is the subject of debate amongst scientists: will global warming alter the precipitation properties of clouds? If more clouds produce more moisture in the lower atmosphere and heavier rainfall from clouds, then their impact will be limited. If, however, the increased number of clouds results in an increase in humidity throughout the troposphere, then global warming will increase because water vapour is the most important greenhouse gas. This issue has yet to be resolved.

COMPUTER MODELLING THE CLIMATE

The first stage in modelling how the climate will behave in the future is to build up a clear picture of how it is behaving at present. This is done by using all the available observations from around the world to calculate the initial conditions at a set of equally-positioned points over the entire surface of the Earth and at various levels throughout the atmosphere, and down into the oceans. Using these starting conditions, the model uses the laws of physics to calculate how the climate will evolve in the future.

For the purposes of modelling, the atmosphere is sliced horizontally into as many as 50 separate levels from a height of 65km (200,000ft) and conditions at each point in each level are calculated from available data

Different sizes of grid are used to obtain different scales of modelling. For example, a detailed grid with a large number of measurement points may be used to gain information about developments in a particular region while a broader grid with fewer measurement points may be used over a wide area in global modelling

Sea-surface conditions are calculated for each point on the grid

Ground measurements are used to calculate conditions at each point on the grid

MONITORING FOR THE FUTURE

THERE ARE STILL MAJOR GAPS IN OUR UNDERSTANDING
OF THE CLIMATE – ADVANCES IN TECHNOLOGY ARE
ESSENTIAL IF WE ARE TO FILL THEM IN.

Scientists and meteorologists are constantly striving to improve technology and current methods of monitoring the climate so that we can gain a better idea of what is going on in the world around us.

THE CONTRIBUTION OF SATELLITES

Weather satellites are a recent development and it was anticipated that their data would illuminate our understanding of climate processes and change. The first priority of satellite programmes is to investigate changes that occur over several years or several decades. This work will concentrate on producing better observations in slowly varying components of the climatic system, such as the global radiation balance, sea surface temperatures and the extent of snow and pack-ice. These observations may hold the key to explaining variations in trends measured by surface instruments. The first analyses of observations made by the microwave sounding units mounted on National Oceanic Atmospheric Administration (NOAA) satellites between 1979 and 1988 suggested it was possible to measure monthly global lower atmospheric temperatures to an accuracy of 0.01°C (0.018°F).

CONFLICTING DATA

Scientists anticipated that satellite data would correlate with existing surface observation from both past and present. However, they soon discovered that the situation is more complicated than was first assumed and the integration of satellite observations into the monitoring of global temperature trends requires great care. The problem with using satellite measurements collected since 1979 is that they show far less warming than surface observations. Indeed, over the 18 years between 1979 and 1997 they show an overall global cooling with no measurable trend in the northern hemisphere and a notable cooling in the southern hemisphere. This is in marked contrast with the surface figures, which show a pronounced warming trend over the same period. Although 18 years is too short a time to draw definite conclusions about longer-term trends, the differences are certainly real and have led to a heated debate within the climatological community. Defenders of the surface observations have

Above: A weather research balloon launched from the National Scientific Balloon Facility near Palestine, Texas. These measure a variety of atmospheric factors up to altitudes of 30km (100,000ft).

Left: A Seasat satellite image of global wind speeds from blue (lowest) through grey and purple to orange (highest).

sought to show that satellite figures are flawed because of changes in the calibration of instruments on different satellites. The satellite researchers on the other hand have stoutly defended their methodology and pointed to the close correlation between their figures and radiosonde balloon data obtained at the same time in the same places.

It is unlikely that instrumental limitations in satellite data are the real cause of the difference. Some of it lies in the fact that microwave instruments measure a thick slice of the lower atmosphere centred on an altitude of around 3km (10,000ft). Indeed the radiosonde evidence shows that, while the Earth's surface has warmed during the 1980s and 1990s, the temperature of the middle troposphere has remained constant. This would help to explain the discrepancy between the two sets of results, but would introduce major complications for climate modellers.

What these differences do underline, however, is that most meteorological observations are made for day-to-day forecasting use and are not well-suited to interpreting long-term shifts in climate change. These limitations cause climatologists untold problems and show that monitoring systems that might play a significant part in swaying decision-making must achieve much more accurate results than they do at present.

GREATER KNOWLEDGE

Obtaining a better understanding of how the climate is changing will continue to depend on making the most effective use of the existing monitoring systems and those that are planned for the future. These will be funded by many different countries, principally to maintain and improve weather forecasting services around the world and so using them to detect subtle changes in the climate will pose the same challenges as in the past.

But new, more ambitious experimental projects, such as the hugely expensive NASA EOS programme, which is having great difficulty getting adequate funding from the American Congress, offer us the prospect of obtaining more comprehensive observations of a wider range of climatic factors. These may include infrared and microwave instruments that are capable of making much more detailed measurements of the temperature structure of the atmosphere, and laser equipment that is able to make better observations of winds. Improved radar systems, such as those flown on the European ERS-1 and 2 satellites, will provide more data about waves and surface winds. However, because these projects are experimental, it will be a long time before they can be used to monitor climate change.

REANALYSIS OF WEATHER FORECASTS

Much of the current monitoring of the climate is used to prepare weather forecasts and this process involves thorough analysis of the available data. The major producers of numerical weather forecasts are now devoting considerable effort to re-examining their output to see what it can tell about the climate: for example, the European Centre for Medium-Range Weather Forecasts (ECMWF) has recently completed a major project covering the period 1979–93. This type of analysis will be able to produce a more complete and up-to-date view of our climate in what looks like becoming a more changeable world.

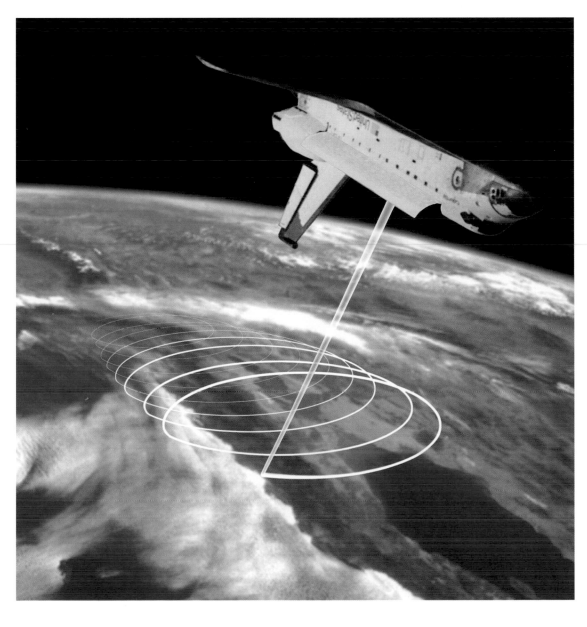

Left. An artist's impression of Sparcle, part of a new generation of measuring devices currently being developed by NASA, which will use laser radar to measure winds from space.

Global warming 184–5

THE COST OF CLIMATE CHANGE

RECENT EXTREME WEATHER HAS DENTED INSURANCE
COMPANIES' PROFITS, PROVIDING A REAL FINANCIAL
INCENTIVE FOR PREDICTING FUTURE CLIMATE CHANGE.

Perhaps the greatest uncertainty about climate change is the issue of whether the weather will become more extreme. Until we know the answer to this question it will not be possible to form a realistic view about the future financial costs of any global warming.

Below: Waves crash onto the coast at Port Lavaca, Texas, USA in September 1961.

ADDING UP THE COST

There is a widely held view that there has been a dramatic upswing in the frequency of extreme weather events in the last decade or two, but this is not supported by the meteorological statistics. What has gone up strikingly is the insurance cost of damage caused by extreme events in the industrialized world. Before 1986 no insured loss had ever exceeded $1 billion, but the damage caused by Hurricane Hugo in 1989 cost five to six times more than this figure and that caused by Hurricane Andrew in 1992 exceeded $15 billion. Much of this rising trend in losses is due to the rapid increase in size and vulnerability of shoreline

communities and a rise in the number of people insuring their properties. Therefore, these figures must be considered carefully because they are caused by a combination of meteorological and economic factors. But with over one-third of the world's population living within 100km (62 miles) of the coast, it is important to ascertain how much, if at all, the meteorological component is contributing to the insurance burden. For this, we need to look at the climatic evidence.

IS THE WEATHER MORE EXTREME NOW?

Careful analysis of the meteorological records kept for Atlantic hurricanes show that storm activity was much greater from 1940–70 than it has been since. In this period, there were more intense storms, stronger winds and longer periods with winds above 120kph (75 mph). The years 1995 and 1996 were both relatively stormy but, on the whole, there have been decreasing frequencies of intense hurricanes since the late 1960s.

In the north-western Pacific, which has more tropical storms than any other region, the annual incidence of hurricanes declined between the late 1950s and the mid-1970s from around 32 to 28, but has returned to over 30 a year in the 1990s. The incidence of intense typhoons has also exhibited a similar dip and rebound.

The same lack of a clear trend applies to mid-latitude storms. In the North Atlantic the incidence of exceptionally intense low-pressure systems reached record high levels at the end of the 1980s and in the early 1990s, but wider analysis provides a more complicated picture. Studies of pressure patterns over the UK and the North Sea since the late 19th century show no evidence that it has been windier at any one period. Measurements of mean wave heights during the winter in the North Atlantic tell a different story: their size increased by some 80 per cent between the early 1950s and the early 1990s, though changes at other times of the year have been much less extreme.

The strength of the westerly winds in the northern hemisphere is best measured by swings in the North Atlantic Oscillation (NAO). When the winds are strong the NAO is said to have a positive value, and it brings warm spells to northern Europe and cold spells to Greenland. When the winds are weak it has a negative value and the climatic situation is reversed. The values of the NAO over the last 130 years show no significant overall trend. Instead there have been periods of stormy winters followed by quieter runs of years. The most striking feature over the last 100 years or so has been the marked dip in values during the 1960s and the rise since then, though the 1990s have witnessed a series of ups and downs.

All this evidence supports the general conclusion reached by the UN Intergovernmental Panel on Climate Change in 1995 that: 'Overall there is no evidence that extreme weather events, or climatic variability, has increased in a global sense through the 20th century, although the data and analyses are poor and not comprehensive. On regional scales there is clear evidence of changes in some extremes and climate variability indicators. Some of these changes have been towards greater variability; some have been towards lower variability.'

FLOODS

This said, flooding is one phenomenon that is definitely becoming more prevalent. Record-breaking floods along the Mississippi in 1993 and 1995, the Rhine in December 1993 and January 1995, the Yangtze in 1995 and 1998, and in central Europe during July 1997 are widely seen as a consequence of global warming. Although an increase in temperature might lead to more heavy rainfall, recent floods may partly be due to the fact that we have changed the hydrology of many river basins by building embankments and reducing the extent of the flood plain. This has accelerated the pace of drainage and made local communities more vulnerable to inundation.

WHAT DOES THE FUTURE HOLD?

Although small changes in the climate could produce large changes in the frequency of extreme events, meteorologists are not sure what will happen. Global warming would lead to more high temperatures and less low ones, with a decrease in daily temperature variability. More interestingly, any warming of the atmosphere and oceans will pump more moisture into the atmosphere, which might increase extreme rainfall and flooding in some areas, combined with more frequent or severe droughts elsewhere in the world.

In the insurance world, these vague estimates of the types of events we may have to face in the future demand more sophisticated forms of risk analysis. When actuaries come to make their risk calculations they have to take not only social and economic factors into account, but also any climatological statistics available. If these figures are not a good guide to the incidence or severity of future events then the insurance industry will find it difficult to cover losses in some extreme events. Therefore, the rising level of insurance losses is generating intense pressure to find ways to improve the analysis of the occurrence of extreme events.

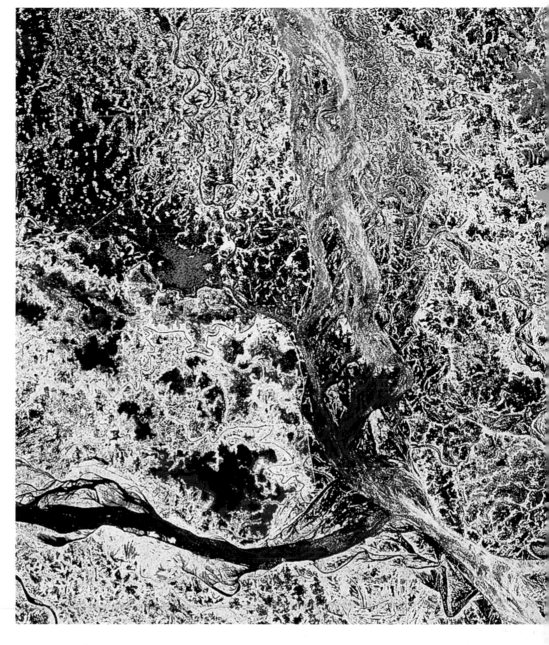

Above: A false-colour image prepared from radar data obtained by the ERS-1 satellite of the floods in Bangladesh in 1993. The image combines observations made on 24 July and 28 August. Areas shown in red were flooded on 24 July and those in blue were flooded on 28 August; those coloured black were flooded on both occasions.

HURRICANE DAMAGE

The maximum potential destruction by hurricanes in the North Atlantic between 1950 and 1998 has been estimated using data based on wind speeds and duration of the storms. This measure provides figures for the destructive power of the hurricanes in any year by combining the number of days when the wind speeds were above a given level with the square of the wind speed, to reflect the fact that most intense hurricanes do by far the most damage. This index shows a marked decline until the early 1990s, as do other measures of hurricanes in the Atlantic.

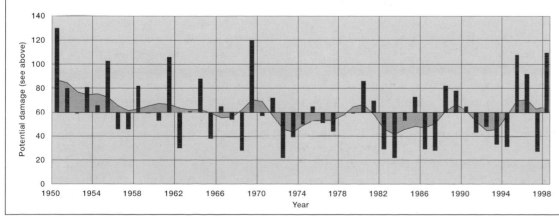

FUTURE FOOD PRODUCTION

THE REPERCUSSIONS OF GLOBAL WARMING MAY AFFECT THE TYPE OF AGRICULTURE PRACTISED AND FUTURE FOOD PRODUCTION ACROSS THE WORLD.

At the moment, we are producing more than enough food to go around. Shortages (particularly in the developing world) are often caused by political complications rather than agricultural shortfalls. With the current rate of population growth, however, we must ensure that we continue to produce enough food in years to come. The future of agriculture will not only be affected by climate change – it also depends on the rising levels of carbon dioxide in the atmosphere, which increase the growth of plants and changes in breeds of plants, including genetic modification.

A GLOBAL MODEL

A study into the future of food production was carried out by Cynthia Rosenzweig, at the Goddard Institute of Space Studies in New York, and Martin Parry, while at the

Below: The surplus crop production in the western world is symbolized by huge stores such as the EEC grain intervention store at Duffield, UK.

Environmental Change Unit in Oxford, UK. They put together a model to show what food production will be like in 2060, based on combined climate predictions from three GCMs that suggest the average temperature will be approximately 4–5°C (7–9°F) higher and carbon dioxide levels will double over the period. They predicted the resulting changes in agricultural yields and then used a world food-trade model to simulate the economic consequences of the climatic changes in terms of food

prices, altered global trade patterns and the number of people who will be at risk of hunger in the developing countries of the world.

They drew on an immense amount of local agricultural expertise – scientists in 18 different countries estimated the potential national grain yields using models based on local production. These agricultural simulations gave an accurate global picture because they were carried out in areas that represent some 70–75 per cent of wheat, maize and soybean production and about 48 per cent of rice production. The scientists combined all these figures with predictions of how crop yields would respond to technical advances over the period under scrutiny. Overall, the conclusion they reached was that by 2060 global grain production will rise to approximately 3300 million tonnes, an increase of 80 per cent on 1990.

WINNERS AND LOSERS

The major finding of this work was a large disparity in the agricultural vulnerability to climate change between the developed and developing countries. Overall, agricultural production will decline only slightly to moderately as a result of global warming, but the impact of this change on different parts of the world will be unequal: yields at high latitudes will increase while those at low latitudes will decline. Near the high-latitude boundaries of production, increased temperature will lengthen growing seasons and the extra warmth in the summer will increase yields. Together with the benefits of extra carbon dioxide, the agricultural gains will be considerable.

In middle and high latitudes, where yields are already high, the extra warmth will shorten the crop-development period and reduce yields. But the increase in heat and the need for more water in these regions will not be very significant, and overall any adverse climatic effects will be outweighed by the benefits of increased carbon dioxide. At lower latitudes, however, the combination of rapid crop development together with rising heat and the need for extra water will lead to significant reductions in output.

Agricultural practices in lower latitudes can be adapted in two ways. First, farmers can respond in terms of crop choices, planting dates and irrigation levels. Second, governments can invest in agricultural equipment and practice to improve matters. Climate change would reduce production by between 11 and 20 per cent, depending on which set of GCM results is used. But, when the responses to change are taken into account, combined with the effects of increased carbon dioxide, the forecast is not as gloomy. The falls in output would only decline by 0–5 per cent if farmers' responses were successful and would range from an increase of 1 per cent to a fall of 2.5 per cent if major adaptations were to take place.

ADAPTING TO CHANGE

In spite of the many uncertainties (for example, variations in regional climate, population growth and how different countries will adapt), these predictions raise interesting

Above: Methods of irrigating arid land, such as this system in the Mojave Desert, California, USA, will have to become more efficient if future climate change is not to pose a growing threat to food supplies.

questions about future world food supplies. At one level, it can be argued that the effect of climate change is small, and that market forces are capable of meeting the challenges ahead. Furthermore, adaptation is the only practical way of improving matters as, even if we are successful in fully achieving the reductions in greenhouse-gas emissions agreed at Kyoto in 1997, and in agreeing further cuts, the impact on food production by 2050 will be tiny. In the meantime, reducing water demand will have a much greater impact on agricultural production, especially in feeding those people living in drought-prone areas who are most at risk from future climate change.

This view may be somewhat misplaced because it understates the destabilizing effects of the growing gap that is emerging between production and needs in the developing world, while countries in the developed world produce ever-greater surpluses of food (see far left). However, this pressure can be regarded as an integral part of continuing negotiations about the future economic order between countries in the developing world and those in the developed world. The part played by agriculture is likely to centre on crop-breeding programmes to increase the crop tolerance for surviving in heat and drought conditions, as well as their ability to ward off pests and diseases. At the same time, changing agricultural practices in order to maximize their contribution to reducing the effects of global warming (see box, right) will become more and more important in the years to come.

REDUCING GREENHOUSE GASES
One aspect of future agricultural development will be how changing practices can alter the amount of carbon and nitrogen that is stored in the soil. Organic methods can, over the years, lead to substantial increases in the carbon and nitrogen levels in the soil. These methods include the reduction of tillage, which means root systems are not disturbed as much, and the application of manure, which enables the soil to retain more organic materials. These techniques are capable of producing yields on a par with modern intensive agriculture, and offer a solution to the major environmental problem of waste from intensive livestock rearing, as well as providing a 'sink' for atmospheric carbon. The potential for agriculture to lock up more carbon in the soil could be an important factor in finding an agreement about future action to reduce the emissions of greenhouse gases into the atmosphere. Each country will have to produce an inventory of its net emissions and so the amount of carbon stored in the ground by organic agricultural practices is likely to play an increasingly important part in international negotiations.

RISING AGRICULTURAL PRODUCTIVITY
The yield of wheat in the UK has tripled over the last 50 years. This rising productivity has been a common feature of agriculture in the developed world, and the developing world is increasingly seeing similar rises in output. It is expected that continuing improvements in breeding crops and in the technology of agriculture and pest and disease management will enable these rises in productivity to continue.

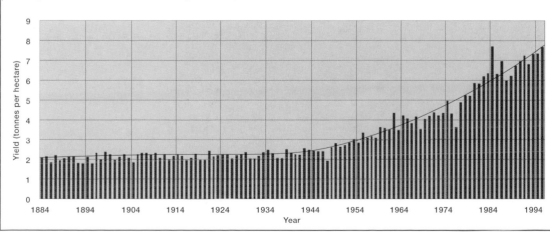

Global warming 184–5

CLIMATE AND HEALTH

CLIMATE CHANGE BRINGS WITH IT A NEW FEAR, NOT OF A METEOROLOGICAL OR ECONOMIC NATURE, BUT SOMETHING MUCH CLOSER TO HOME – OUR HEALTH.

In recent years, concerns have grown about the impact that climate change might have on human health For the majority of people in the industrialized world, this would centre around air pollution and skin cancer but the potential spread of infectious diseases is also alarming.

CONSEQUENCES OF GLOBAL WARMING

A major assessment of the effects of climate change on human health carried out by the World Health Organization concluded that the most dangerous consequences of global warming for human health would be the increase in heatwaves, droughts, storms and floods, which may grow in both frequency and intensity as average temperatures and precipitation rise. Nearly half the world's population lives in cities and the proportion is rising, and the effects of heatwaves and air pollution will be most strongly felt in urban areas. Current models predict that, by the middle of the 21st century, many major cities around the world could be seeing several thousand extra heat-related deaths every year. Changes in the production of some types of air pollutants (for example, the fine particulates given off by burning diesel) will also increase the incidence of respiratory disorders, such as hay fever and possibly asthma.

Any rise in sea level would have a number of adverse effects, including population displacement, loss of farmland, salt-water intrusion, contamination of water supplies, changes in the distribution of diseases (see below) and increases in death and injury due to flooding.

Other social–demographic disruptions may include shortages of fresh water, and stresses on agricultural systems and other resources.

Increased exposure to ultraviolet radiation is likely to cause a further increase in the incidence of skin cancer, ocular lesions such as cataracts and a weakening of the immune system. It is thought, surprisingly, that the popularity of sun-bathing will be only partially cancelled out by increased awareness of the dangers involved and the widespread use of high-protection-factor sun lotions. Any continued decline in ozone levels in the upper atmosphere will make matters worse. However, most of the decline takes place at high latitudes in the spring and has less of an effect on the top sun-bathing locations at low latitudes in the summer, where if anything ozone levels are increasing.

BUGS AND VIRUSES

Recently, there have been major changes in the distribution of malaria, dengue fever and encephalitis (mosquito- and tick-borne), as well as outbreaks of plague and hantavirus pulmonary syndrome (both rodent-borne) and Lyme disease (tick-borne). There has been a marked resurgence in malaria over the last 20 years, with regional warming and increased rainfall thought to be responsible for an increase in the breeding of infected mosquitoes and their migration, which spreads malaria into adjacent non-endemic areas. In the next century, global warming is expected to increase the incidence of malaria by 50–80 million more cases each year. Dengue viruses are also carried by mosquitoes. Higher temperatures not only increase insect numbers but also their biting frequency and the percentage that are infectious. In addition, cholera has spread because flood-borne sewage triggers algal bloom in tropical coastal areas – the algae that bear the cholera pathogen then contaminate drinking water.

Right: Sinking a deep tubewell in Bangladesh in order to obtain safe water supplies.

THE GREATER PICTURE

Climate change will have some beneficial effects on human health. In cool temperate countries, milder winters should reduce the mortality peak among older people at this time of year. Similarly, in the hottest areas, any increase in summer temperatures might actually kill mosquito populations, which will cut down the spread of infectious diseases. Nevertheless, the overall direct health consequences of global warming are reckoned to be adverse. Many of the anticipated factors affecting human health will have greatest impact in developing countries at low latitudes, where people may already be vulnerable because of shifts in agricultural production.

Any predictions of future health consequences of a warmer climate do, however, have to be put into the context of the progress that has been made during the last century in public health and of the major challenges that are currently facing the developing world. The most pressing public-health issues concern infant mortality caused by the lack of clean, safe water supplies and adequate nutrition. These problems must be confronted, whether or not the climate warms as it is predicted to do. As such, together with maintaining past progress in controlling the spread of infectious and contagious

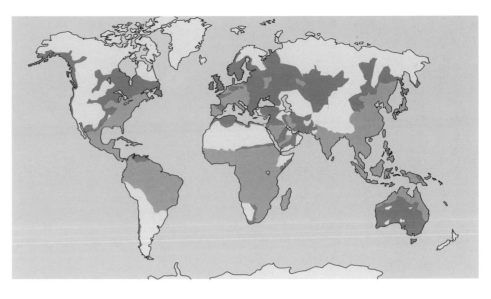

diseases, these challenges represent a continuing and massive task for health services in the developing world. In this context, any changes caused by shifts in climate, such as more frequent droughts or floods, are likely only to be a small part of what will be one of the greatest concerns facing humankind.

MALARIA DISTRIBUTION

Present extent of Malaria

Projected extent of Malaria in 2055

At present, malaria mainly affects people living in tropical and subtropical areas of the world. Some computer models predict that, with the rise in temperature that is expected over the next 50 years, malaria could start to threaten many more regions.

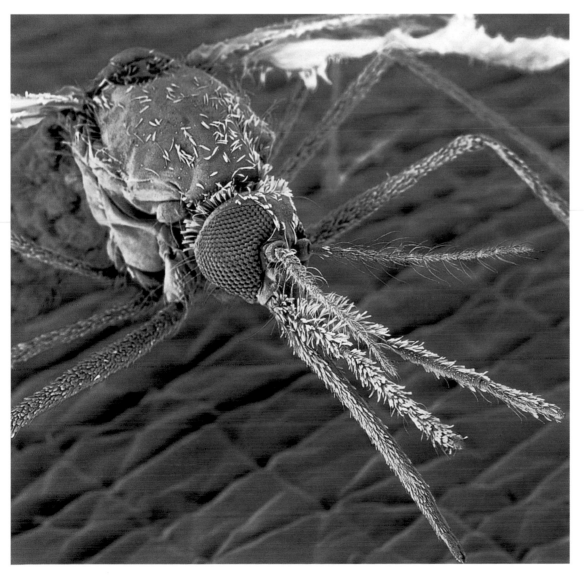

Left: A scanning electron micrograph of the head of a malaria-causing mosquito. The piercing proboscis below the pair of mandibular palps is used by the female of the species to draw blood. The parasite that causes malaria is transmitted in the saliva of female mosquitoes.

Sea-level rise 182–3

Global warming 184–5

WEATHER MODIFICATION

FROM RAIN DANCES TO CLOUD SEEDING, HUMANS
HAVE ALWAYS TRIED TO CONTROL AND MODIFY THE
CLIMATE, SOMETIMES WITH DEVASTATING RESULTS.

*Above: Hopi Indians in ceremonial
clothing and masks dance to the rain
god in Shungopavi, Arizona, USA.*

Attempts to modify the weather provide a number of
important messages about both our limited capacity
to understand how the atmosphere behaves and to
anticipate the consequences of interfering with something
we do not fully comprehend.

EARLY EFFORTS

Humans have always tried to influence the weather by the
power of prayer. Perhaps the best-known example is the
Hopi Rain Dance in Arizona, which has an extraordinary
reputation for producing the late-summer rainfall needed
to bring maize to ripeness. A more dramatic story tells of
a severe drought in the 19th century when Mexican villagers
formally announced that after eight days they would stop
praying, then eight days later they would destroy the church
and eight days after that they would kill the priest and nuns
if it did not rain. Fortunately for all concerned, torrential rain
came after just four days.

The first formal proposal for weather modification was
put forward by pioneering American meteorologist James
Pollard Espy in 1841. He wanted to light large bonfires
every 32km (20 miles) along a 1100km (700 mile) front west

of Mississippi once a week to create rainfall across the
eastern USA. This grandiose scheme never received any
serious support but various proposals involving explosion
and combustion devices were dreamed up throughout the
rest of the 19th century. There are various legends about
American 'Rainmakers' who produced floods that could
not be stopped and who were then either sued or driven
from town by the outraged citizens.

The one combustion device that did work was called
Fog Investigation Dispersal Operation (FIDO). This was a
brute-force approach used in England during World War II
to enable bombers to take off and land in fog. It involved
burning large quantities of oil along the sides of the runways
to lift the fog. It was effective but hugely expensive and
polluting, and could only be justified in a wartime situation.

CLOUD SEEDING

The first scientific efforts to modify clouds occurred soon
after World War II. By then an accepted theory of
precipitation had been developed, which stated that the
essential first stage was the formation of ice crystals on
dust and other particles in the atmosphere (condensation
nuclei). So, it was argued, introducing more condensation
nuclei in clouds should increase rainfall. The first
experiments to establish whether this was the case were
conducted by Vincent Schaefer and Bernard Vonnegut,
working at the General Electric Research Laboratory, USA,
in 1946 using dry ice and silver iodide.

The apparent success of these experiments led to a rush
to judgement. Within a few years operational programmes,
mainly to suppress hail, were being conducted in France,
Italy, Kenya, the Soviet Union, Switzerland and the USA.
Yet many meteorologists were suspicious of the claimed
successes and it slowly emerged that the results did not
live up to early expectations. A little knowledge can be a
dangerous thing – the first attempt to seed a hurricane to
increase precipitation and drain its energy led to it suddenly
veering off course and slamming into Savannah, Georgia.

The fundamental limitation to the initial work was that
nobody could be certain what might have happened
without seeding. Only 'double-blind' statistical trials can
provide the answer: this means that neither the people
seeding the clouds nor those making the measurements
of precipitation know whether or not the flares contained
seeding material. This is essential to ensure there is no bias
in selecting clouds or interpreting their precipitation
patterns. Only when all the measurements have been made
is the identity of the seeding and non-seeding flares
revealed. The results of such rigorous tests were non-
conclusive. In some cases there was evidence that some
increase in rainfall could be achieved in the right conditions.
Conversely, Project WHITETOP to seed summertime
cumulus in Missouri demonstrated that seeding could
actually reduce precipitation by up to 50 per cent.

Cloud seeding then got a bad name because projects
associated with disastrous events coincidentally took place
nearby. For example, the South Dakota School of Mines

was working near the Black Hills of Dakota in 1972, when a flash flood burst a dam and swept away much of Rapid City, killing over 200 people. In 1974, the US Department of Defence admitted to a clandestine programme that took place from 1967–72 to enhance summer-monsoon rainfall over Vietnam. Then in the late 1970s the American Project Stormfury, aiming to modify hurricanes in the tropical Atlantic and Caribbean, ran into international opposition as Mexico expressed concern about the impact of operations on rainfall in northern parts of the country, which depend on tropical storms for much of their precipitation.

WEIGHING UP THE BENEFITS

Although work continues on research into cloud seeding, both to understand the physics and to develop practical schemes, the general meteorological view is sceptical of its potential. Perhaps the most productive area of weather modification lies in trying to reduce the occurrence of damaging hail and French wine growers claim to be having success in this regard.

There is a benefit in maintaining a sense of realism about how much control we can exercise over the weather or climate. The weather-modification boom was part of a wave of technological optimism in the 1960s, which also proposed blocking the Bering Straits, diverting Russian rivers, controlling the Gulf Stream off Florida and covering the Arctic ice with soot, all on the basis that these measures would 'improve' the climate.

Similarly, we should apply a moderate approach when advocating drastic action to prevent the predicted impact of human activities. If we are not certain that any specific course of action will achieve a given outcome then we must concentrate on doing those things that look like a good idea anyway, such as conserving energy and improving public transport, in the hope that they will contribute to the objective of minimizing our impact on the climate while improving our lives generally.

Above: A rocket being launched in Moldavia in an attempt to reduce the impact of damaging hail.

Left: Operation FIDO (Fog Investigation Dispersal Operation). A Royal Air Force Lancaster bomber is shown taking off between two bands of flame, which were able to disperse fog successfully.

SEA-LEVEL RISE

OF ALL THE IMPACTS OF GLOBAL WARMING, THE
PROSPECT OF THE SEA LEVEL RISING AROUND US IS
POTENTIALLY THE MOST ALARMING.

Above: The Thames barrage in London, UK, which is raised when there is a danger of flooding and which has been used effectively on a considerable number of occasions since it was opened in 1983.

Right: Pack-ice breaking off the Trinity Peninsula in Antarctica.

Dramatic images of London and New York submerged beneath the rising waters are a media favourite when presenting the threat of rising temperatures.

CAUSES OF SEA-LEVEL RISE

Although the concept of global sea levels rising appears simple, it is not easy to measure in practice. The worldwide change in sea level is known as eustasy, and reflects not only thermal expansion and contraction of the oceans, but also fluctuations in the amount of water stored in the ground, in lakes and inland seas, glaciers, ice caps and the ice sheets of Greenland and Antarctica. We also have to consider the much slower shifts caused by the tectonic movement of the Earth's surface. In addition, sea-level measurements must be taken from several geographical locations so that the effect of local land movements can be ruled out.

Over the course of the 20th century, sea levels are thought to have risen by approximately 18cm (7in), but that figure could be as low as 10cm (4in) or as high as 25cm (10in). Oceanic thermal expansion is a direct consequence of the global warming that has taken place over the last 100 years: estimates based on atmosphere–ocean computer models suggest that this warming may be responsible for 2–6 cm (⅘–2⅕in) of the rise. Melting glaciers and ice caps caused some 4cm (1½in).

Other potential contributors to the rising sea level include changes in the mass of the Antarctic and Greenland ice sheets, extraction of groundwater for agricultural purposes and changes in the volume of inland lakes and seas. Even after all these factors have been considered, we can only

explain less than half of the observed sea-level rise and this uncertainty shows how difficult it is to be confident about any projections of future rises.

FUTURE RISES

The speed of future sea-level rises depends on how the climate responds to global warming and, in particular, whether precipitation rates at high latitudes increase.

If warming leads to more precipitation over Antarctica and Greenland, then any additional melting of the ice sheets is likely to be more than balanced by the accumulation of extra snowfall at higher levels. At lower latitudes, where glaciers and ice caps have been receding since the late 19th century, warming has outweighed any increase in rainfall. So, for much of the 21st century the rise in sea level will be dominated not only by the thermal expansion of the oceans but also by the melting of ice caps and glaciers at lower latitudes. If the predicted rise in temperatures continues beyond 2100, then the melting of the ice sheets of Greenland and Antarctica is likely to become a problem. The latest cautious estimate is that sea levels will rise by 27cm (11in) by 2100 but it may range from 17 to 49cm (7 to 19in).

Any rise in sea level will affect different parts of the world in different ways. This is because the Earth's crust is still rising as a result of the melting of the great ice sheets,

which covered much of the northern hemisphere during the last Ice age. This movement means that the impact of sea-level rise will often vary from one coastal site to another. In addition, local factors including groundwater extraction and land reclamation further complicate matters. Therefore, accurate analysis of the impact of a rise in sea level on specific parts of the world will depend on a combination of an improved global model taking into account the movement of the Earth's crust and a better understanding of local geology.

All of this suggests that in the foreseeable future our response to sea-level rises will rely on the same planning methods that are used at present: by keeping an eye on the coastal areas that are known to be vulnerable and link this knowledge with the available weather data regarding temperature and rainfall, meteorologists can produce estimates of how certain types of severe weather – when combined with rising tides – will increase the incidence of damaging water levels in the future. Then decisions will have to be made about whether building adequate protection is a worthwhile investment, or whether in some circumstances it is better to allow the sea to encroach.

COLLAPSE OF THE ICE SHEETS

Most of the world's ice sheets seem to be relatively stable, but what would happen if one of them were suddenly to collapse? The West Antarctic ice sheet is particularly vulnerable because it rests on a rockbed well below sea

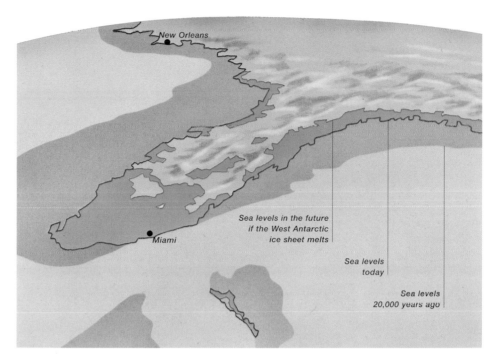

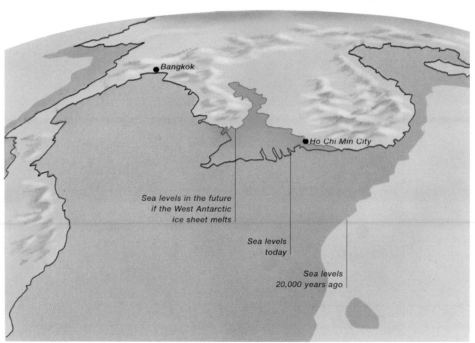

level that could, in theory, break away at any time. If this were to happen, the consequence would be a rapid rise in sea levels of several metres worldwide.

This possibility has been the subject of a large number of studies – the Ross Ice Shelf, which abuts onto the West Antarctic ice sheet, acting as a dam, may disintegrate over the next 200 years and as a result the larger sheet may itself collapse over the following 20–100 years. This would lead to an increase in sea levels of 60–120cm (23–47in) per century. However, until global temperatures have risen by at least 8°C (14°F), snowfall will continue to build up on the ice sheets, which will actually reduce sea levels. As temperatures will not reach this level until 2200 at the earliest, this potential disaster looks a fairly distant prospect.

FLORIDA AND SOUTH-EAST ASIA
During the last Ice age, some 20,000 years ago, sea levels in Florida and South-East Asia were 120m (390ft) lower than they are today. If the West Antarctic ice sheet does melt, then sea levels will rise dramatically posing many problems to coastal areas.

Global warming 184–5

THE NEXT 100 YEARS

Global warming is one of the most important climatic issues for the next 100 years. But is the climate really going to carry on getting warmer and, if so, can we do anything to stop it?

There is broad scientific agreement that the rise in global temperatures in recent decades is partially due to the fact that, since the Industrial Revolution, the amount of carbon dioxide in the Earth's atmosphere has risen by approximately 25 per cent. Past and present figures have been used in computer models to predict how much warmer the climate will become over the next century or so. It is thought that, by the year 2100, global temperatures will be 1.5–4.5°C (2.7–8.1°F) warmer, depending of course on a wide variety of economic and political factors.

Modelling global warming

One challenge in using computers to model global warming is accurately reproducing the warming that has already taken place over the last 100 years or so. In particular, they must explain why the warming that has occurred during the 20th century (and its temporary abatement in the 1940s and 1970s) has not tallied directly with increases in the build-up of greenhouse gases in the atmosphere. Although the burning of fossil fuels produces carbon dioxide, which contributes to the greenhouse effect, another by-product of this combustion is sulphate aerosols, which have the opposite effect: they have a cooling influence on the climate because they reflect sunlight back out to space rather than trapping it within the Earth's atmosphere as carbon dioxide does.

The Meteorological Office Hadley Centre in the UK has explored the consequences of doubling the amount of carbon dioxide in the atmosphere, using advanced computer models that reproduce the behaviour of and interaction between oceans and atmosphere. Scientists there conducted three experiments, which started with conditions as they were in 1860 and predicted them up to 2050. In the first model, the control, the carbon dioxide levels were kept constant. In the second, the carbon dioxide was increased gradually to represent the impact of all greenhouse gases; and, in the third, the combined effects of carbon dioxide and sulphate aerosols were considered. The results were illuminating: by 2050, if the carbon dioxide levels double, the sulphate aerosols will counterbalance the effects, especially over India, China, Mexico and southern Africa, where global warming will noticeably slow down.

These results reflect what has been happening during the 20th century, with the sulphates holding down the temperature rise. This model also predicts a less

Left: Destructive hurricanes – such as Hurricane Mitch, which hit this banana plantation in Honduras at the end of October 1998 – are predicted to become a more frequent occurence in the 21st century.

dramatic warming of 2.5°C (4.5°F) over the period 1860–2050, and this warming is reduced to 1.7°C (3°F) with the effect of sulphates.

How will this affect the weather?

General warming trends are much easier to predict than effects at ground level in terms of the weather we experience. According to computer models, polar regions are expected to warm more than equatorial regions, reducing the temperature gradient and, hence, the amount of energy transported polewards. This could reduce the strength of mid-latitude depressions in winter and move the principal storm tracks to slightly higher latitudes. Rising sea surface temperatures in the tropical oceans could also lead to more frequent and intense tropical cyclones. Recent trends do not support these predictions. Furthermore there is a worrying tendency for the treatment of various climatic features (for example, the properties of clouds, sulphate particulates or snow cover) in models to produce contradictory regional perturbations. So there is no agreed view of how global warming will alter basic climatic features such as the strength of winter mid-latitude westerlies, the Indian monsoon or the incidence of tropical cyclones.

Planning for the future

Overall, there is still huge uncertainty about what we can actually do to minimize the impact of human activities on the climate (see right). A better understanding of climatic processes and change would help us to plan our lives more effectively, from choosing a holiday destination to being better informed about the latest government decision on taxing fuel. It would also help us to appreciate how we should tackle the threat of global warming. Perhaps we have found the right balance by addressing climate change through issues that touch on health, education and welfare. But how will we know when to change the balance? Only a clearer appreciation of what is happening to the climate can ensure that we get our priorities right in the years to come.

This is the challenge for the 21st century. The best we can hope is that global warming will follow the lower rather than higher predictions and that the climate will be no more variable than in the past. But we need to anticipate developments in good time and not simply indulge in a knee-jerk reaction when new extremes hit us. Whatever the future holds, we are all in the same boat and so concerted international responses must be central to any strategy. Understanding the climate's complex interactions and interpreting the wider implications of regional extremes are vital to reaching an agreement on the sacrifices that we may all have to make in the future.

Global warming has been the subject of several conferences (Rio de Janeiro in 1992, Berlin in 1995 and Kyoto in 1997), but achieving lasting reductions in greenhouse-gas emissions will involve more than just setting targets. Trading in emission quotas between developed and developing countries may play a part: each country would be entitled to emit a certain quantity of gases and developing countries could sell their quotas to industrial nations so that the overall emission would not exceed permitted levels. The going will get tougher as the economically and politically acceptable options are used up. Advanced nations may be willing to introduce energy-conservation measures, improve public transport and close down ageing power stations but, when it comes to imposing taxes on energy consumption or restricting the use of private vehicles, their resolve to meet targets may weaken. Similarly, developing countries, whose energy consumption per capita is lower, may be reluctant to curb economic growth.

GLOSSARY

References in *italic* are to other entries within the glossary.

Aerosols Minute particles of liquid or gas that are suspended in the atmosphere. Dust, sulphur dioxide, sea salt and carbon are examples of particles that can be suspended in this way. Aerosols are important as the site for the *condensation* of water droplets and ice crystals, and as participants in various atmospheric chemical reactions. Aerosols, such as those formed from volcanic eruptions, can lead to a cooling at the Earth's surface.

Air mass A large body of air that has virtually consistent properties, such as temperature and moisture distribution, across its horizontal extent.

Albedo The reflectivity of a surface or body. Snow, for example, has a high albedo as it reflects most of the solar radiation it receives.

Anticyclone Part of the atmosphere where the atmospheric pressure increases towards the centre – often called a 'high'. Winds in such a system circulate in a clockwise direction in the northern hemisphere and in an anticlockwise direction in the southern hemisphere. Anticyclones bring calm clear weather. (See also *depression*.)

Aurorae Luminous curtains or streamers of light seen in the night sky at high latitudes. These occur electrically-charged particles from the Sun are guided by the Earth's *magnetic field* to the polar regions where they collide with atoms in the upper atmosphere.

Biosphere The parts of the Earth's surface and atmosphere that are occupied by life forms, on land and in the air and oceans.

Blocking A phenomenon caused by stationary *anticyclones*, usually in northern hemisphere mid-latitudes, that prevents the normal movement of pressure systems, resulting in periods of abnormal weather.

Chlorofluorocarbons (CFCs) A family of inert, non-toxic and easily liquefied chemicals, used as coolants in refrigerators and air-conditioners, as propellants in aerosol cans and as solvents. They are implicated in both *ozone* layer depletion and *global warming*.

Cirrus Wispy clouds that form at high altitudes.

Climate The long-term statistical average of weather conditions. Global climate represents the long-term behaviour of such parameters as temperature, *air pressure*, *precipitation*, soil moisture, runoff, cloudiness, storm activity, winds and ocean currents, integrated over the full surface of the globe. Regional climates, analogously, are the long-term averages for geographically limited domains on the Earth's surface.

Cold front The boundary line between advancing cold air and a mass of warm air under which the cold air pushes like a wedge. Cold fronts are often associated with heavy rainfall. (See also *warm front*.)

Condensation The conversion of a vapour or gas into a liquid.

Continental drift The lateral movement of continents as a result of sea-floor spreading. (See also *plate tectonics*.)

Convection A type of heat transfer whereby hot, less dense air rises and is replaced by cooler, denser air.

Coriolis force A deflecting force affecting moving objects in a rotational system. A body moving across the rotating Earth is deflected to the right of the path of motion in the northern hemisphere and to the left in the southern hemisphere. The force is weakest around the equator. It is important in the formation of *cyclones*.

Cryosphere The portion of the Earth's *climate* system that comprises the world's ice masses, sea ice, glaciers and all snow deposits.

Cumulonimbus A cloud that forms at low altitude, with a flat base and flowering top. This type of cloud can often be seen during thunderstorms.

Cumulus Detached clouds, generally dense with sharp outlines, that develop vertically in the form of rising mounds, domes and towers – the rising upper part often resembles a cauliflower.

Cyclone The name given to a *hurricane* that occurs in the Bay of Bengal.

Depression Part of the atmosphere where the atmospheric pressure decreases towards the centre – often called a 'low'. Winds in such a system circulate in an anticlockwise direction in the northern hemisphere and in a clockwise direction in the southern hemisphere. Depressions bring unsettled weather. (See also *anticyclone*.)

Dew-point The temperature at which moist air becomes saturated and deposits dew. Above the ground, *condensation* of vapour into water droplets takes place when the dew-point is reached.

Easterlies Winds that blow from the east.

Ecosystem A conceptual view of a plant and animal community, emphasizing the interactions between the living and non-living parts, and the flow of materials and energy between these parts.

Electromagnetic radiation The emission and propagation of energy in the form of varying electric and magnetic fields, which can travel through a vacuum. Classes of electromagnetic radiation are defined by their wavelengths, from gamma-rays through to radiowaves.

El Niño The warm phase of the *ENSO* when sea surface temperatures in the equatorial Pacific rise above normal.

El Niño Southern Oscillation (ENSO) A reversal of atmospheric pressure patterns that takes place every few years across the tropical Pacific causing above or below normal sea surface temperatures which result in changes in rainfall distribution throughout the tropics.

Equinox The two dates, one occurring in spring and the other in autumn, when the day and night are of equal length. (See also *solstice*.)

Eustasy The worldwide global changes in sea level caused by changes in water volume due to the formation and melting of ice sheets, changes in the temperature of the water or variations in the volume of ocean basins induced by changing ocean ridges. (See also *plate tectonics*.)

Evaporation The conversion of a liquid into a gas, the rate of which increases with temperature.

Evapotranspiration The transfer of water from vegetated land surfaces into the atmosphere by *evaporation* and *transpiration*.

Fog Minute water droplets (with radii of 1–10 *micrometres*) suspended in the atmosphere which reduce the visibility to below 1km (0.6 miles).

Gaia hypothesis This hypothesis holds that living organisms on Earth (including micro-organisms) actively regulate atmospheric composition and *climate*, helping provide stability in the face of challenges such as the increasing luminosity of the Sun or increasing *greenhouse-gas* emissions.

General Circulation Models (GCMs) A computational model or representation of the Earth's *climate* used to forecast changes in *climate* or weather.

Geostationary satellite A satellite with an orbit of 24 hours, travelling in the same direction as the Earth is rotating – this means that it remains constantly at the same position over the equator, typically at an altitude of 35,900km (22,300 miles).

Glacial epoch See *ice age*.

Global warming An overall increase in the Earth's temperature.

Greenhouse effect An atmospheric process in which an increase in the concentration of atmospheric trace gases (greenhouse gases) leads to a decreases the amount of *radiation* that escapes directly into space from the lower atmosphere, and consequently a rise in the temperature throughout the Earth's atmosphere. Short-wave solar *radiation* can pass through the atmosphere relatively unimpeded, but long-wave terrestrial *radiation*, emitted by the warm surface of the Earth, is partially absorbed and then re-emitted by these trace gases, causing an increase in temperatures.

Groundwater Water that occupies holes and crevices in rocks as a result of rain percolating into the ground.

Hadley Cell An atmospheric circulation pattern in the tropics: moist warm air rises near the equator, spreads out to the north and south and sinks at about 20–30°.

High-pressure system See *anticyclone*.

Holocene The relatively warm epoch, which started around 10,000 years ago and runs up to present time. It is marked by several short-lived particularly warm periods, the most significant of which, some 6000 years ago, is called the Holocene optimum.

Humidity The amount of water vapour in the atmosphere. (See also *relative humidity*.)

Hurricane A storm with particularly violent winds, consisting of a region of intense low pressure surrounded by revolving winds of very high speeds. Known as hurricanes in the West Indies and Gulf of Mexico, they are referred to as *cyclones* and *typhoons* when they occur elsewhere in the world.

Hydrological cycle See *water cycle*.

Ice age A period during the history of the Earth when there were larger ice sheets and mountain glaciers than today. The most recent ice age, the Pleistocene, encompassed much of the last 2.5 million years. Overall, ice ages comprise only 5–10 per cent of all geological time. During major ice ages, great ice sheets formed in high latitudes and spread out to cover as much as 40 per cent of the Earth's land surface. See also *interglacial*.

Insolation The solar radiation (from INcoming SOLar radiATION) received at any particular area of the Earth's surface, which varies from region to region depending on latitude and weather.

Interglacial A warmer period during an *ice age* when the major ice sheets receded to higher latitudes.

Intertropical Convergence Zone (ITCZ) A narrow low-latitude zone in which air masses originating in the northern and southern hemispheres converge and often produce cloudy, showery weather.

Ion An atom or molecule with an electric charge due to the loss or gain of electrons.

Ionosphere A region of the upper atmosphere above 85km (50 miles) in which an appreciable concentration of ions and free electrons exist. This region is also known as the thermosphere.

Isobar A line on a weather chart that links points of equal atmospheric pressure.

Isotherm A line on a weather chart that links points of equal temperature.

Isotopes Atoms of a single element with the same number of protons but different numbers of neutrons, and therefore different atomic weights. Isotopes are labelled with the atomic weight preceding the symbol of the element (eg ^{18}O denotes an oxygen isotope with an atomic mass of 18, instead of the usual 16 (^{16}O).

Jet streams Strong winds that occur in the upper *troposphere* – they have a complex interdependent relationship with weather systems at lower levels because they each affect the course of the other.

Katabatic wind A wind formed when very cold air builds up in upland areas and becomes dense enough to slide downhill. This is a major weather feature in some places, eg the Mistral in the Mediterranean.

Kelvin waves Wave patterns that occur in both the atmosphere and the oceans. These relatively fast-moving waves move eastwards across the Pacific and, together with *Rossby waves*, affect the depth of the *thermocline* and hence El Niño.

La Niña The cold phase of the *ENSO* when sea surface temperatures in the equatorial Pacific drop below normal.

Lapse rate The rate at which temperature falls as altitude increases.

Latent heat of vaporization The heat absorbed during the change of state from a liquid to a vapour.

Low-pressure system See *depression*.

Magnetic field The force field around a magnet. The Earth has such a field, as its liquid metal core acts like a magnet.

Magnetosphere The area of influence of the *magnetic field*.

Mesosphere A region of the upper atmosphere above the *stratosphere* that extends from 50 to 85km (30 to 50 miles) above the surface.

Microclimate The *climate* of a small area, sometimes surrounding organisms or in specific habitats, which is different to the normal climate of the surrounding area.

Micrometre (µm) 1×10^{-6}m (or about 0.00004 inches).

Monsoon A seasonal reversal of winds, which blow inshore during the summer bringing heavy rains and blow offshore during the winter bringing dry weather. It is of greatest meteorological importance in southern Asia.

Noctilucent clouds Beautifully coloured clouds which sometimes occur at altitudes of 75–90km (45–55 miles). They are visible around midnight at latitudes above 50° when they reflect light from the Sun below the horizon.

North Atlantic Oscillation (NAO) A pattern of atmospheric circulation in the North Atlantic which is measured in terms of the difference in pressure between Europe and Greenland. In winter, there are strong westerly winds that bring mild weather to Europe and cold weather to Greenland. In the summer, the winds follow a more meandering pattern, which bring mild weather to Greenland and cold weather to Europe.

Ozone A molecule made up of three atoms of oxygen (O_3), which is found throughout the atmosphere. In the *stratosphere*, it provides a protective layer shielding the earth from ultraviolet *radiation* and subsequent harmful health effects on humans and the environment. Lower down, in the *troposphere*, it is a major component of photochemical smog.

Permafrost Permanently frozen soil found in Arctic, tundra, taiga and some mountain regions.

Photosynthesis The process by which plants use sunlight to convert carbon dioxide and water from the air into energy. Vital for maintaining an optimum atmospheric composition, this process absorbs carbon dioxide from and emits oxygen into the atmosphere.

Phytoplankton Microscopic marine organisms (mostly algae and diatoms) which are responsible for most of the *photosynthesis* in the oceans.

Plate tectonics The theory that the Earth's crust is divided into several large moving plates that act as rigid slabs floating on a viscous liquid mantle. It is used to explain features of the Earth's surface such as mountains and oceanic trenches and processes such as earthquakes and volcanoes.

Prairies Large open areas of grassland in North America.

Precipitation Moisture that falls on the Earth's surface from clouds in the form of rain, hail or snow.

Pressure High pressure, see *anticyclone*; low pressure, see *depression*.

Proxy data Any sources of information that contain indirect evidence of past changes in the weather (for example, tree rings, ice cores and ocean sediments).

Radar Radiowaves or microwaves used to gauge the distance of objects by measuring the time taken for a *radiation* pulse to travel from the transmitter to the object and back to an adjacent receiver.

Radiation See *electromagnetic radiation*

Radio-sonde A free-floating balloon carrying instruments that measure temperature, *pressure* and *humidity* as it rises through the atmosphere.

Relative humidity The amount of water vapour in the air, expressed as a percentage of the maximum it is able to hold at that temperature before becoming saturated. (See also *humidity*.)

Rossby waves Wave patterns that occur in both the atmosphere and the oceans. The relatively slow-moving Rossby waves move westwards across the Pacific and (together with *Kelvin waves*) affect the depth of the *thermocline* and El Niño.

Solstice The longest and shortest days of the year. (See also *equinox*.)

Squall line A line of intense *convective* activity, often over 100km (60 miles) in length – there can be thunderstorms and strong winds at low levels in a squall line.

Steppes Huge areas of flat treeless grassland in south-east Europe and Siberia.

Stevenson shelter A standard device for housing ground-level meteorological instruments, designed to ensure that accurate shade temperature measurements can be made.

Stratocumulus Clouds that form in a low layer of disjointed grey masses.

Stratosphere A region of the upper atmosphere, which extends from the *tropopause* to about 50km (30 miles) above the Earth's surface. The *ozone* layer is located in the stratosphere.

Stratus Clouds that form a continuous grey layer over large areas.

Sunspots Dark blotches on the Sun's surface indicating increased solar activity.

Synoptic forecasting Weather forecasting based on the analysis of various factors (*pressure*, temperature and wind speed) recorded at a specific time over a big area.

Taiga Coniferous forest found at high northern latitudes between the *tundra* and *steppes* of Siberia.

Temperate regions Areas in the mid-latitudes that are characterized by mild weather.

Thermocline A region of the ocean between the warm upper layer of water and the colder deeper water, over which there is a rapid change in temperature.

Thermohaline circulation The deep-water circulation of the oceans that is driven by differences in density due to variations in salinity and temperature.

Thermosphere See *ionosphere*.

Tornado An intensely destructive, advancing whirlwind formed from the strongly ascending air currents associated with vigorous thunderstorms.

Trade winds Persistent winds that blow from the north-east in the northern hemisphere and from the south-east in the southern hemisphere between the equator and latitudes of 30°N and S.

Transpiration The process by which water is drawn from the soil into plant roots, transported through the plant, and then *evaporated* from leaves and other plant surfaces into the air.

Troposphere The lower atmosphere, from the ground to an altitude of about 8km (5 miles) at the poles, about 12km (7.5 miles) in mid-latitudes, and about 16km (10 miles) in the tropics. Clouds and weather systems, as experienced by people on the ground, take place within the troposphere.

Tropopause The boundary between the *troposphere* and the *stratosphere*.

Tundra The name derived from the Finnish word that describes the vast treeless regions from northern Scandinavia to the Bering Sea and across Alaska and northern Canada.

Typhoon The name for a *hurricane* that occurs in the western Pacific Ocean.

Warm front The boundary line between advancing warm air and a mass of colder air over which it rises. Warm fronts are often associated with the arrival of warm calm weather. (See also *cold front*.)

Water cycle The process by which water passes between the oceans, atmosphere and land. This involves the processes of *precipitation, evaporation* and *transpiration*.

Water table The level below which fissures and pores in the ground are saturated with water.

Westerlies Winds blowing from the west.

INDEX

ACKNOWLEDGEMENTS

ABBREVIATIONS:

c – centre; b – bottom; t – top;
l – left; r – right

AKG – AKG, London; AMS – AMS Brooks Library, USA; AP – The Associated Press Ltd; AU – Auscape International; B2C – Biosphere 2 Centre, Inc; BAL – Bridgeman Art Library; BC – Bruce Coleman Ltd; BCA – Bryan & Cherry Alexander; BL – Billie Love Historical Collection; BW – Bradford Washburn; CG – Commission for Glaciology, Bavarian Academy of Sciences, Munich; CO – Corbis UK Ltd; COL – Colorific; CPL – Cephas Picture Library; E – Explorer; EI – Environmental Images; FLPA – Frank Lane Picture Agency; GSFC – Goddard Space Flight Centre; HG – Hulton Getty Picture Collection; IWM – Imperial War Museum; JT – Judy Todd; MC – Mountain Camera; MIT – MIT Museum, Cambridge, MA; MSFC – Marshall Space Flight Centre; N – Novosti; NASA – National Aeronautics and Space Administration; NML – National Meteorological Library & Archive; OSF – Oxford Scientific Films; PE – Planet Earth Pictures; RF – Rex Features; RH – Robert Harding Picture Library; SeaWiFS – SeaWiFS Project, NASA/Goddard Space Flight Centre and ORBIMAGE; SP – Still Pictures; SPL – Science Photo Library; TS – Tony Stone Images; WFA – Werner Forman Archive

PHOTOGRAPHIC ACKNOWLEDGEMENTS

Front and back endpapers TS/Art Wolfe, 2–3 FLPA/Fred Hoogervorst/Foto Natura, 4–5 OSF/NASA, 6b HG, 7 SPL/Martin Bond, 8cr MC, 8bl E/L.Bertrand, 8br RF/SIPA, 9tl PE/Bourseiller/I&V, 9tr PE/Bourseiller/ I&V, 10–11 SPL/NASA, 14b NASA, 16l SPL/Los Alamos National Laboratory, 17t SPL/Los Alamos National Laboratory, 19tr FLPA/Mark Newman, 21t RH/C Black, 21b RH/Rolf Richardson, 22 PE/Bourseiller/I&V, 23t PE/Bourseiller/I&V, 23bl NASA/GSFC, 24 NASA, 25tr AKG/Erich Lessing, 25c SPL/NASA, 25b SPL/Hale Observatories, 26t SPL/Library of Congress, 26b SPL/Keith Kent, 27tl SPL/Frank Zullo, 28tl CO, 29t SeaWiFS, 29bl SPL, 29br FLPA/Leo Batten, 30–1 SPL/European Space Agency, 32b SPL/Peter Menzel, 33tl RH, 33tr S D Burt, 33b SP/Mark Edwards, 34 SPL/European Space Agency, 36 NASA, 38 E/ L Bertrand, 39t CO, 39b RF/Sipa Press, 40cl SP/Ed Parker, 40br SPL/David Scharf, 41 SPL/David Nunuk, 42t SPL, 42b RH, 46–7b SPL/David Parker, 47t SPL/N, 47tr CO, 48–9 RH/Art Wolfe, 50b BCCA, 50–1t BCA, 51t HG, 51b HG, 52–3t BCA, 53b SPL/US Geological Survey, 54b SPL/NASA, 55tl SPL/Pekka Parviainen, 56t BCA, 56b RH/Stuart Wallace, 57t SPL/US Geological Survey,

58 BCA, 59t BCA, 59b WFA, 60tl SPL/NASA/GSFC, 60tr SPL/NASA/GSFC, 60cr SPL/NASA/GSFC, 60bl SPL/NASA/GSFC, 62b TS/Kim Westerskov, 62–3t RH, 63b NASA, 64 OSF/Francois Valla/Okapia, 65tl SPL/Csiro/Simon Fraser, 65tr FLPA/ S McCutcheon, 66–7 BCA, 68 FLPA/Roger Tidman, 69 BCA, 70 BCA, 71 FLPA/Roger Tidman, 72l OSF/Richard Packwood, 72tr SPL/Simon Fraser, 73t FLPA/J Stoick/Dembinsky, 73b FLPA/ T Leeson/Sunset, 74l BCA, 74br FLPA/M Newman, 75 OSF/Andrew Park, 76 BCA, 77 CPL/Walter Geiersperger, 78 FLPA/ S McCutcheon, 79tl CO, 79tr FLPA/ S McCutcheon, 80–1 BW/ Panopticoa Gallery, 82b MC, 83t MC, 84 SPL/ W Bacon, 85 AP, 86tl RH/Dave Willis, 87 JT, 88 BCA, 89 CO, 90 RF/Sipa, 91 RH/ P Wysocki, 92t CG, 92b CG, 93 TS/RGK Photography, 94–5 CPL/Herve Champollion, 96b OSF/Stuart Bebb, 97 SPL/Peter Menzel, 97t CPL/Beatty, 97b OSF/Alain Christof, 99t EI/Ian Richards, 99b CPL/Mick Rock, 100t PE/Rosemary Calvert FRPS, 100–1b BC/Charlie Ott, 101t FLPA/T&P Gardner, 102 CPL/Juan Espi, 103t SPL/Adam Hart-Davis, 104 BAL/National Maritime Museum, London, 105tl BAL/Castello Del Buonconsiglio, Trent, 105bl RF, 106–7 SPL/John Heseltine, 108b SPL/University of Dundee, 109t FLPA/Mark Newman, 111br AP, 112 SP/Hjalte Tin, 113 SPL/John Heseltine, 114 SP/Julio Etchart, 115t FLPA/P Moulu/ Sunset, 116 CO, 117tr AMS, 117b SP/Thomas Raupach, 119b SP/Mark Carwardine, 119tr SPL/Dr Richard Legeckis, 120tl BL, 120b AKG/Erich Lessing, 121t BAL/Giraudon, 122–3 SPL/E R Degginger, 124b SPL/Jim Gipe/Agstock, 125 SPL/JVZ, 126tl FLPA/M Newman, 126cl, SPL/National Centre for Atmospheric Research, 126bl SPL/NCAR, 127t SP/Oliver Langrand, 128b OSF/Warren Faidley, 129t AP/ Hutchinson News/Mark Colson, 130t FLPA/Roger Wilmshurst, 130b FLPA/Lee Rue, 131t OSF/CM Perrins, 132 AKG, 133 CO, 134–5 FLPA/Fred Hoogervorst/ Foto Natura, 136br SP/Max Fulcher, 137tl OSF/Stephen Dalton, 137tr PE/Carol Farneti, 138tl SP/Gil Moti, 139tr SPL, 139 br NOAA/National Climatic Data Centre, 141 EI/Doug Perrine, 142 SPL/George Bernard, 143t SP/Yves Lefevrre, 143b SP/Klein/Hubert, 144 SPL/NRSC LTD, 145 SPL/George Holton, 146 NASA/ GSFC/Scientific Visualisation Studio, Greg Shirah 147 COL/David Portnoy/Blackstar, 149tl SPL/NASA, 149tc SPL/NASA, 149tr NML/K Woodley, 149cl SPL/NASA, 149cr SPL/NASA, 149b FLPA/David Hosking, 150–1 OSF/Michael Fogden, 152br SP/Mathieu Laboureur, 153t AU/Jeff Foot, 153br SPL/US Geological Survey, 154b SP/Gordon Wiltsie, 155t SP/Gil Moti, 155b SP/D Escartin, 156cl SPL/Hank Morgan, 156br SP/Harfmut Schwarzbach, 157 SPL/John Mead, 158 AU/Jean-Paul

Ferrero, 159t SP/Chris Caldicott, 159b SP/Klein/Hubert, 160 SP/Adrian Arbib, 161 FLPA/David Hosking, 162 SP/Mark Edwards, 163tr SP/Hjalte Tin, 164 FLPA/ F Hoogervorst/Foto Natura, 165t SPL/NASA/GSFC, 165bl FLPA/W Wisniewski, 165br CO, 166 PE, 167l SP/Klaus Andrews, 167tr FLPA/David T Crewcock, 168–9 B2C/Gonzalo Arcila, 170tl MIT, 172l NASA, 172r NASA, 173 NASA/MSFC, 174 C/Bettman, 175t SPL/ESA, Eurimage, 176t EI/Patrick Sutherland, 176b OSF/Donald Specker, 177 OSF/Mary Cordano, 178 OSF/Paul McCullagh, 179b SPL/Eye of Science, 180 CO, 181t N/B Kapnin, 181b IWM, 182tl FLPA/David Warren, 182–3bc SPL/CNES, 1989 Distribution Spot Image, 184 AP, 185 SPL/Simon Fraser.

ARTWORK ACKNOWLEDGEMENTS

12l Cottier & Sidaway, 12–13c Raymond Turvey, 13tr Raymond Turvey, 14 Cottier & Sidaway, 15 Raymond Turvey, 18–19 Chris Forsey, 20 Chris Forsey, 25 Cottier & Sidaway using data from NOAA, 27 Raymond Turvey, 35 Raymond Turvey, 38 Cottier & Sidaway using data from the University of Berne, 43 Raymond Turvey, 44 Raymond Turvey, 45 Raymond Turvey, 50tl Rudi Vizi, 50cl Cottier & Sidaway, 51 Kenny Grant, 52tl Rudi Vizi, 52cl Cottier & Sidaway, 55 Raymond Turvey, 57 Chris Forsey, 65 Cottier & Sidaway using data from the World Data Centre, 68tl Rudi Vizi, 68cl Cottier & Sidaway, 71 Raymond Turvey, 76 Cottier & Sidaway using data from Rutgers University, 79 Rudi Vizi, 82tl Rudi Vizi, 82cl Cottier & Sidaway, 93 Chris Forsey, 96tl Rudi Vizi, 96cl Cottier & Sidaway, 98 Rudi Vizi, 103 Cottier & Sidaway using data from the Journal of Interdisciplinary History (1980), 108tl Rudi Vizi, 108cl Cottier & Sidaway, 109 Rudi Vizi, 110–11 Raymond Turvey, 111tr Raymond Turvey, 115bc Chris Forsey, 115br Cottier & Sidaway, 117 Raymond Turvey, 118 Raymond Turvey, 119 Raymond Turvey, 121 Cottier & Sidaway using data from the University of Massachusetts, 124tl Rudi Vizi, 124cl Cottier & Sidaway, 127 Rudi Vizi, 128 Rudi Vizi, 129 Rudi Vizi, 136t Rudi Vizi, 136cl Cottier & Sidaway, 137 Raymond Turvey, 138 Raymond Turvey, 139 Cottier & Sidaway using data from the Indian Institute of Tropical Meteorology, 140 Raymond Turvey, 141 Raymond Turvey, 148cl Cottier & Sidaway using data from the University of East Anglia, 148b Cottier & Sidaway using data from NOAA, 152tl Rudi Vizi, 152cl Cottier & Sidaway, 171 Raymond Turvey – based on artwork from ECMWF, 175 Cottier & Sidaway using data from the University of Colorado, 177 Cottier & Sidaway, 179 Kenny Grant, 183 Chris Forsey.

AUTHOR'S ACKNOWLEDGEMENTS

Because this book draws on lengthy personal involvement in climate matters, it is difficult to identify all the people who have helped me to form a view on the many facets of climate, how it has changed and its impact on all our lives. Among the meteorological community I would like to thank Chris Folland, David Parker, Tony Slingo, John Mitchell, Bruce Callendar, Jack Hopkins at the UK Meteorological Office, David Anderson, Tim Palmer, Tony Hollingsworth, Peter Jannsen, Anders Peerson, Austen Woods, at the European Center for Medium Range Weather Forecasting, Grant Bigg, Keith Briffa, Tom Holt, and Mike Hulme and Phil Jones at the University of East Anglia, David Cotton at the University of Southampton, John Harries at Imperial College, Dominic Reeve at Sir William Halcrow and Partners Ltd, Christopher Landsea at NOAA Hurricane Research Division, Miami, David Robinson at Rutgers University, Tom Karl at NOAA Climate Data Center, John Christy at University of Alabama, Jan Lindstrom at the University of Helsinki, Franz Fliri and Norman Lynagh for helpful discussions, and the provision of data and other material, which in one way or another, was essential for the completion of this book. In addition, I am particularly grateful to Richard Alley, Roger Barry, Per Gloersen, Michael Hambrey, Sari Kovats, Michael Mann, Martin Parry, Julia Slingo, Jon Snow, and Kevin Trenberth for helpful advice in the initial planning of this book.